From Polynomials
to Sums of Squares

FROM POLYNOMIALS
TO SUMS OF SQUARES

Terence Jackson

Department of Mathematics
University of York

Institute of Physics Publishing
Bristol and Philadelphia

British Library Cataloguing-in-Publication Data

A catalogue record for this book is available from the British Library.

ISBN   0 7503 0364 6   hardback
           0 7503 0329 8   paperback
           0 7503 0365 4   site licence

Library of Congress Cataloging-in-Publication Data are available

Published by Institute of Physics Publishing, wholly owned by The Institute of Physics, London

Institute of Physics Publishing, Techno House, Redcliffe Way, Bristol BS1 6NX, UK

US Editorial Office: Institute of Physics Publishing, The Public Ledger Building, Suite 1035, 150 South Independence Mall West, Philadelphia, PA 19106, USA

Typeset in TeX using the IOP Bookmaker Macros
Printed in the UK by J W Arrowsmith Ltd, Bristol BS3 2NT

⟩ Contents

〉 Preface

This book describes a journey through some of the beautiful foothills of algebra and number theory based around the central theme of factorization. It begins with very basic knowledge of rational polynomials, gradually introduces other integral domains, and eventually arrives at sums of squares of integers. I have tried throughout the main text to make the treatment very concrete. So, for instance, when discussing polynomials the emphasis is very much on practice with the arithmetic of polynomials. Similarly, the algebra of quadratic domains is continually illustrated by referring to specific quadratic integers. Statements about general integral domains are mentioned where appropriate but are met in an abstract context only in the appendices.

A potential drawback of using numerical examples with polynomials or quadratic integers is that the arithmetic is much more tedious than say arithmetic among the ordinary rational numbers. This is where a suitable program running on a small computer can be an ideal assistant, since it can handle the computational details while letting the reader direct the calculations. Many general purpose computer algebra systems could be used for this purpose but they tend to be expensive and to need some knowledge of their special syntax. The programs created to accompany this book run on IBM PCs and are intended to be inexpensive and very easy to use. I have assumed that the user will know how to start the programs running on their own system, but once started they contain all necessary instructions for their use. The programs and text can each be used alone but are designed to enhance each other when used together. Throughout the text there are practical activities mostly involving a computer. I have deliberately tried to make these straightforward to encourage the reader. More theoretical exercises are left to the end of each chapter.

Very little is needed in the way of mathematical prerequisites. The main requirements are familiarity with complex numbers and with some elementary number theory including congruences and the finite field $\mathbb{Z}_p$. In order to make the book as self-contained as possible many background results have been given in appendices. They can be taken on trust at first and only consulted if the reader wishes to have more detail.

⟩ Software copyright and site licence

The 'From Polynomials to Sums of Squares' software package is protected by copyright. The copyright is owned by IOP Publishing Ltd and will be defended against violations of copyright legislation. Users are permitted (1) to use the software on a single computer, (2) to make one back-up copy of the software or (3) to install the software on a single hard disk provided the original copy is kept solely for back-up or archival purposes.

Warning
The User acknowledges that software in general is not defect- or error-free and may become contaminated resulting in damage to other software and equipment. The User is advised to take all such steps as are usual to verify the integrity of the software and/or to minimize the risk of damage or loss arising from such defects, errors or contamination such as taking out insurance against such a risk. The User's attention is specifically drawn to the following paragraphs.

Exclusion of warranties/liability
This software and documentation is sold 'as is' without warranty as to performance, merchantibility or fitness for any particular purpose. All other warranties, whether statutory, implied or otherwise, are also hereby excluded.

Site licence
A site licence is available for institutional purchasers of the software. Purchasers of the site licence are licensed (subject to registration) to make multiple copies of the software for use solely within the purchasing institution, or to load the software into a network confined to the institution. To purchase the site licence please use the order form overleaf.

'FROM POLYNOMIALS TO SUMS OF SQUARES' SOFTWARE SITE LICENCE ORDER FORM

Please check the Institute of Physics Publishing catalogue for current site licence price before ordering.

I wish to purchase a site licence for the 'FROM POLYNOMIALS TO SUMS OF SQUARES' software package (ISBN 0 7503 0365 4)

Customer details:

Name: . . . .

Address: . . . .

. . . .

. . . .

. . . .

Method of payment:

Credit card: Visa ☐   AmEx ☐   Mastercard/Access ☐
Card no: . . . .
Expiry date:
Today's date:
Signature: . . . .

Cheque ☐   Bank draft ☐   International money order ☐
Please send me a Pro-forma Invoice ☐

Send order to:
UK/Rest of World: Customer Services Department
                   Institute of Physics Publishing
                   Techno House, Redcliffe Way, Bristol BS1 6NX, UK
USA/Canada: Institute of Physics Publishing
                   c/o AIDC, 2 Wintersport Lane
                   PO Box 20, Williston, VT 05495-0020, USA

⟩ Chapter 1

⟩ Polynomials in one variable

⟩1.1 Polynomials with rational coefficients

Usually when we begin to study algebra at school we first meet and work with expressions containing a letter, like $x - 1$ or $4x^2 - 9x + 2$, before we learn that such expressions are called 'polynomials' †. We commonly think of x as just standing in place of a number; so that in $4x^2 - 9x + 2$ for instance we could replace x by 5 and find that the value of the polynomial is then $4*5^2 - 9*5 + 2 = 57$. We could actually write out 'four times the square of a given number minus nine times the given number plus two' but that is more difficult to remember, and work with, than the shorthand $4x^2 - 9x + 2$. Medieval mathematicians often did write out expressions in words; only they wrote in Latin. Similarly the value of $x^2 + \frac{2}{3}x - 1$ when x is replaced by 5 is $5^2 + \frac{2}{3}*5 - 1 = 27\frac{1}{3}$. Note that

$$(4*5^2 - 9*5 + 2) + (5^2 + \tfrac{2}{3}*5 - 1) = 5*5^2 - 8\tfrac{1}{3}*5 + 1 = 84\tfrac{1}{3}$$

so that the value of $5x^2 - 8\frac{1}{3}x + 1$ when $x = 5$ is $84\frac{1}{3}$. We could express this by putting

$$(5x^2 - 8\tfrac{1}{3}x + 1)_{x=5} = (4x^2 - 9x + 2)_{x=5} + (x^2 + \tfrac{2}{3}x - 1)_{x=5}$$

and likewise if x were replaced by 3 say we would have

$$(5x^2 - 8\tfrac{1}{3}x + 1)_{x=3} = (4x^2 - 9x + 2)_{x=3} + (x^2 + \tfrac{2}{3}x - 1)_{x=3}.$$

† We could of course have polynomials involving any other letter, such as $b + 2$, $y^3 - 1$ or $3z^2 + z + 7$; but for simplicity I shall try to use x throughout.

An analogous equation would also hold if, in each of $5x^2 - 8\frac{1}{3}x + 1$, $4x^2 - 9x + 2$ and $x^2 + \frac{2}{3}x - 1$, we replaced x by any other number instead of 3 or 5. Of course it gets very tedious to write down separate and closely similar equations such as

$$(5x^2 - 8\tfrac{1}{3}x + 1)_{x=7} = (4x^2 - 9x + 2)_{x=7} + (x^2 + \tfrac{2}{3}x - 1)_{x=7}$$

$$(5x^2 - 8\tfrac{1}{3}x + 1)_{x=-2} = (4x^2 - 9x + 2)_{x=-2} + (x^2 + \tfrac{2}{3}x - 1)_{x=-2}.$$

We could conveniently summarise all such equations by writing

$$5x^2 - 8\tfrac{1}{3}x + 1 = (4x^2 - 9x + 2) + (x^2 + \tfrac{2}{3}x - 1) \qquad (1.1)$$

meaning that when we replace x throughout (1.1) by any fixed number we obtain a true statement about numbers. In a similar way we could write

$$3x^2 - 9\tfrac{2}{3}x + 3 = (4x^2 - 9x + 2) - (x^2 + \tfrac{2}{3}x - 1) \qquad (1.2)$$

meaning again that when we replace x everywhere in (1.2) by a fixed number we obtain a true statement about numbers. For example, if we choose $x = -11$, $(3x^2 - 9\frac{2}{3}x + 3)_{x=-11} = 472\frac{1}{3}$, $(4x^2 - 9x + 2)_{x=-11} = 585$, $(x^2 + \frac{2}{3}x - 1)_{x=-11} = 112\frac{2}{3}$ and indeed $472\frac{1}{3} = 585 - 112\frac{2}{3}$ in agreement with (1.2).

However we can also expand our mental world considerably by regarding x in (1.1) and (1.2) as a genuinely new object to adjoin to our existing number system and we can do this in a way consistent with our usual rules for addition and multiplication. We first agree that whenever we multiply x by itself we shall just call the result x^2. In general we define the different powers‡ of x as

$$x^0 = 1, x^1 = x, x^2 = x * x, x^3 = x * x^2 = x^2 * x = x * x * x, \qquad (1.3)$$

and for each natural number k

$$x^{k+1} = x * x^k = x^k * x = x * \ldots * x. \qquad (1.4)$$

We write the sum of x^k and x^k as $2x^k$, $x^k + 2x^k$ as $3x^k$, and $x^k + \ldots + x^k$ as nx^k (where there are n terms). We also write x^k itself as $1x^k$. Then for any natural numbers a and b we have

$$ax^k + bx^k = (a + b)x^k. \qquad (1.5)$$

‡ In a term of the form x^k we call k the *exponent* of x or the *degree* of the term.

In order to include terms in our polynomial algebra such as the $-8\frac{1}{3}x$ in (1.1) we shall actually allow the multiplying numbers a and b above to be any rational numbers, whether positive negative or zero. Of course when we write the product of a number and a power of x, such as $(7/6) * x^2$ we usually cannot simplify it further, as we could for instance if we had $(7/6) * 9$ and simplified this to $21/2$. We have to think of $(7/6) * x^2$ as just representing a pairing of $7/6$ with x^2. There are however three cases in which we can effect such a simplification. We have already mentioned $1x^k$ as just meaning x^k. The product $(-1)x^k$ is also written more conveniently as $-x^k$ and for any k we agree to interpret $0x^k$ as 0. Then the equation (1.5) still makes sense if either a or b is negative or zero.

Example 1.1
For any number a we have

$$ax^k + 0x^k = (a + 0)x^k = ax^k \tag{1.6}$$

which is just what we would expect with $0x^k = 0$. Also

$$1x^k + (-1)x^k = (1 + (-1))x^k = 0x^k = 0 \tag{1.7}$$

and more generally

$$bx^k + (-b)x^k = 0. \tag{1.8}$$

So adding $(-b)x^k$ is the same as subtracting bx^k.

We can express a sum of different powers or multiples of powers of x such as $5x^7$, $-13\frac{2}{3}x^3$, $6x$, 2 by simply writing

$$5x^7 - 13\frac{2}{3}x^3 + 6x + 2.$$

In this instance the term containing the largest exponent is $5x^7$. We call 7 the *degree* of the whole polynomial, and the numbers 5, $-13\frac{2}{3}$, 6, 2 the *coefficients* of the powers of x. In general if the term of highest degree in a polynomial is say $a_k x^k$, for some coefficient $a_k \neq 0$, then any other non-zero term will have a smaller exponent and so will be of one of the forms $a_{k-1}x^{k-1}$, $\ldots$ $a_1 x$, a_0 $(= a_0 x^0)$. So if the polynomial is called f or $f(x)$ it can be represented as

$$f(x) = a_k x^k + a_{k-1}x^{k-1} + \ldots + a_0 \tag{1.9}$$

where $a_k \neq 0$ but some or all of the other coefficients may well be zero. Note that ordinary numbers themselves can be regarded as polynomials in which x does not occur at all. They are called *constants* or *constant polynomials* and non-zero numbers are then polynomials of degree 0. We do not assign a degree to the number 0. If

$$g(x) = b_n x^n + b_{n-1} x^{n-1} + \ldots + b_0 \tag{1.10}$$

is another polynomial with degree n we may choose the names f, g so that $k \geqslant n$. Then as in (1.1), (1.2) and (1.5) we write†

$$f(x) + g(x) = a_k x^k + a_{k-1} x^{k-1} + \ldots + a_{n+1} x^{n+1} \tag{1.11}$$
$$+ (a_n + b_n) x^n + \ldots + (a_0 + b_0)$$

and

$$f(x) - g(x) = a_k x^k + a_{k-1} x^{k-1} + \ldots + a_{n+1} x^{n+1} \tag{1.12}$$
$$+ (a_n - b_n) x^n + \ldots + (a_0 - b_0).$$

Activity 1.1 Run the accompanying algebra programs and select option 1, 'Polynomial Arithmetic', from the main menu. Then choose 'rational polynomial arithmetic' and try entering the polynomial $2x^3 + 1$. Notice that you enter polynomials as you would write them on paper. That is, in this case, you first press '2', then 'x', then '3', then '+' and finally '1' followed by 'ENTER'. The computer recognizes that the '3' is meant to be an exponent and automatically writes it as a superscript. It also labels the polynomial as 'f'. Now enter the polynomial $x^2 + 1$, which will be labelled as 'g'. The computer will find the sum or difference of f and g, and will call the result 'h', if you type '$f + g$' or '$f - g$' (and then 'ENTER'). What is $f + g$, $f - g$, $g - f$?

Activity 1.2 You can alter either of the polynomials f or g by entering for example '$f = 2/3x^4 - 6x^3 + 5$' or '$g = 3x^5 - 12x^3 + 2$'. Alter one or both of the polynomials f and g as appropriate to work out
(i) $(2x^3 + 1) + (3x^5 - 12x^3 + 2)$;

† The convention used in (1.11) and (1.12) is that the initial terms $a_k x^k, \ldots, a_{n+1} x^{n+1}$ are omitted if $k = n$.

(ii) $(2/3x^4 - 6x^3 + 5) - (3x^5 - 12x^3 + 2)$;
(iii) $(2/3x^4 - 6x^3 + 5) + (3x^5 - 12x^3 + 2)$.

The computer will also understand if you write, for example, '$f + x^2$' or '$g = h - 7x^3 + 2$'. This is useful if you just want to alter one of the displayed polynomials slightly and do not need to use the additional terms repeatedly.

Addition of polynomials satisfies the following five important properties:

A1. Any two polynomials can be added together to produce a third polynomial.

A2. Addition of polynomials is *associative* : that is, if f, g and h are polynomials then

$$(f + g) + h = f + (g + h).$$

A3. Addition of polynomials is *commutative* : if f and g are polynomials then

$$f + g = g + f.$$

A4. There is a zero polynomial 0 so that $f + 0 = f$ for any polynomial f.

A5. If f is any polynomial there is another polynomial, written $-f$, (and called the *additive inverse* of f), such that

$$f + (-f) = 0.$$

Properties A1–A5 express the fact that polynomials form an Abelian group with respect to addition (see Appendix 1). The zero polynomial mentioned in property A4 is just the ordinary number 0; and in property A5 the polynomial $-f$ is obtained from f by reversing the signs of all the coefficients of f. Being able to find additive inverses is really the same as being able to subtract any polynomial from another. Subtracting g from f for instance is the same as finding a polynomial h such that $f = g + h$; and if we add $-g$, the additive inverse of g, to each side, we get $(-g) + f = (-g) + g + h = 0 + h = h$. So adding $-g$ amounts to subtracting g.

Example 1.2
If g is the polynomial $x^7 - 6x^3 + 4$ then the additive inverse of g is $-x^7 + 6x^3 - 4$ since $(x^7 - 6x^3 + 4) + (-x^7 + 6x^3 - 4) = 0$.

By extension of (1.3) and (1.4) x^{i+j} means $x * x \ldots * x$ for $i+j$ terms, which amounts to i $x - terms$ followed by j $x - terms$; or in other words $x^{i+j} = x^i * x^j$. Similarly we define the product of $a_i x^i$ and $b_j x^j$ by

$$(a_i x^i) * (b_j x^j) = a_i b_j x^{i+j} \tag{1.13}$$

and the product of the two polynomials f and g, in (1.9) and (1.10) above, is then

$$f(x) * g(x) = \sum_{i=0}^{k} \sum_{j=0}^{n} a_i b_j x^{i+j}. \tag{1.14}$$

If f and g are each non-zero polynomials, the term of highest degree in fg is $a_k b_n x^{k+n}$, so that

the degree of a product fg is the sum of the degrees of f and g. (1.15)

When written out in full the expression on the right of (1.14) is

$a_k b_n x^{k+n} + a_k b_{n-1} x^{k+n-1} + \ldots + a_k b_0 x^k + a_{k-1} b_n x^{k-1+n} + \ldots + a_{k-1} b_0 x^{k-1} + \ldots + a_0 b_n x^n + \ldots + a_0 b_0$

and when like powers are collected together this is

$a_k b_n x^{k+n} + (a_k b_{n-1} + a_{k-1} b_n) x^{k+n-1} + (a_k b_{n-2} + a_{k-1} b_{n-1} + a_{k-2} b_n) x^{k+n-2} + \ldots + (a_1 b_0 + a_0 b_1) x + a_0 b_0$

which can be written as

$$f(x)g(x) = \sum_{r=0}^{k+n} \sum_{i=0}^{r} a_i b_{r-i} x^r \tag{1.16}$$

provided that any coefficients $a_{k+1}, a_{k+2}, \ldots, b_{n+1}, b_{n+2}, \ldots$ are interpreted as zero.

Activity 1.3 Enter the polynomial $2x^4 - 3x^3 + 6$. If you already have polynomials labelled f and g then this one will be called 'h'. In order to give it the name f you must enter '$f = h$'; or alternatively you could have entered '$f = 2x^4 - 3x^3 + 6$' at the beginning. By altering g as appropriate, and then typing '$f * g$', find the products of the following polynomials: (i) $2x^4 - 3x^3 + 6$ and $3x^5 - 12x^3 + 2$; (ii) $2x^4 - 3x^3 + 6$ and x; (iii) $2x^4 - 3x^3 + 6$ and $x + 1$. Leaving g as $x + 1$, how can you go on to calculate $(2x^4 - 3x^3 + 6) * (x + 1)^2$; $(2x^4 - 3x^3 + 6) * (x + 1)^3$; …?

By now you will have noticed that once the polynomials f and g have been decided upon at the beginning of the session they stay the same until you deliberately decide to change them by, for example, writing '$f = \ldots$'. On the other hand, the result of any $+$, $-$ or $*$ operation always appears in the 'h' register, so if you want to preserve the value of 'h' you must move the current h to another position by typing for instance '$g = h$' before using these operations again.

Several of the properties of multiplication are similar to those of polynomial addition.

M1. Any two polymomials can be multiplied together to produce a third polynomial.

M2. Multiplication of polynomials is *associative*: that is, if f, g and h are polynomials then

$$(f * g) * h = f * (g * h).$$

M3. Multiplication of polynomials is *commutative* : if f and g are polynomials then

$$f * g = g * f.$$

M4. There is an *identity* polynomial 1 (the ordinary number 1) so that $f * 1 = 1$ for any polynomial f.

The analogue of property A5 for addition would hold if each polynomial f had a *multiplicative inverse*, which would be a polynomial g such that $f * g = 1$. This is not true unless f is non-zero and is of degree 0 (or in other words is a non-zero number). However there is a weaker property which does hold.

M5. If h_1, h_2 and f are polynomials with $f \neq 0$ then

$$h_1 f = h_2 f \text{ implies that } h_1 = h_2.$$

This property enables us to cancel a common polynomial factor from both sides of an equation and so is called a *cancellation* law. There is also a property which ties addition and multiplication together.

AM. If f, g and h are any polynomials then

$$f * h + g * h = (f + g) * h.$$

This says that we can expand out brackets with polynomial expressions just as we are used to doing with numbers. Property AM is a

generalization of (1.5) and is called the *distributive* law. Properties A1–
A5, M1–M5 and AM say that this polynomial algebra is an integral
domain (see Appendix 1) and it is denoted by $\mathbb{Q}[x]$.

Example 1.3
If one of the multiplying polynomials is a constant then (1.14) shows
that we multiply a polynomial by a number c just by multiplying every
coefficient by c:

$$c * (a_k x^k + a_{k-1} x^{k-1} + \ldots + a_0) = c a_k x^k + c a_{k-1} x^{k-1} + \ldots + c a_0. \quad (1.17)$$

Thus

$$0 * (a_k x^k + a_{k-1} x^{k-1} + \ldots + a_0) = 0 x^k + 0 x^{k-1} + \ldots + 0 = 0;$$

and

$$(-1) * (a_k x^k + a_{k-1} x^{k-1} + \ldots + a_0) = -a_k x^k - a_{k-1} x^{k-1} - \ldots - a_0$$

which is the polynomial's additive inverse.

⟩1.2 Polynomials with coefficients in $\mathbb{Z}_p$

So far we have been using polynomials whose coefficients are rational
numbers. For a given prime p we can work just as easily with
polynomials all of whose coefficients come from $\mathbb{Z}_p$, the field of residues
modulo p (see Appendix 1). The set of all such polynomials is written
$\mathbb{Z}_p[x]$ and the basic operations of addition, subtraction and multiplication
in $\mathbb{Z}_p[x]$ are defined exactly as in (1.11)–(1.16). So all the properties
A1–A5, M1–M5 and AM hold as before, and $\mathbb{Z}_p[x]$ is also an integral
domain.

Activity 1.4 Make sure that you have selected the polynomial arithmetic
option from the main menu and then hold down the 'Ctrl' key and, while
doing so, tap the 'P' key. The computer will ask you to choose a prime
so, to begin with, type '7' (and then press 'ENTER'). If you already have
some polynomials displayed their coefficients will be reduced modulo 7.
Now enter the polynomial $17x^3 - 5x^2 + 3/4$. It will be displayed as
$3x^3 + 2x^2 - 1$. Clearly $17 \equiv 3 \pmod 7$ and $-5 \equiv 2 \pmod 7$ but why
has $3/4$ been replaced by -1? This is because, modulo 7, $3 \equiv 4 * (-1)$;

just as, in rational arithmetic, $3/4 = 0.75$ because $3 = 4 * 0.75$. The computer writes every coefficient as an integer chosen, in this case, to lie between -3 and 3. If it is working in $\mathbb{Z}_p[x]$ it uses the complete set of residues consisting of the integers i with $-p/2 < i \leqslant p/2$.

Activity 1.5 Simplify the following expressions in $\mathbb{Z}_7[x]$: (i) $(3x^3 + 2x^2 - 1) * (2x + 1)$; (ii) $(3x^3 + 2x^2 - 1)^2$; (iii) $(x - 1) * (x - 2) * (x - 3) * (x - 4) * (x - 5) * (x - 6)$.

There is a close and far reaching connection between arithmetic in $\mathbb{Z}_p[x]$ and arithmetic in $\mathbb{Q}[x]$. For example, in $\mathbb{Q}[x]$, we have

$$(x^2 + 5x - 1) * (x^3 - 7x^2 + 9x - 2) = x^5 - 2x^4$$
$$- 27x^3 + 50x^2 - 19x + 2 \tag{1.18}$$

and if we now reduce all the coefficients modulo 3 we have

$$(x^2 - x - 1) * (x^3 - x^2 + 1) \equiv x^5 + x^4 - x^2 - x - 1 \pmod{3}$$

which implies

$$(x^2 - x - 1) * (x^3 - x^2 + 1) = x^5 + x^4 - x^2$$
$$- x - 1 \text{ in } \mathbb{Z}_3[x]. \tag{1.19}$$

Similarly (1.18) immediately implies that

$$(x^2 - 1) * (x^3 - 2x^2 - x - 2) = x^5 - 2x^4$$
$$- 2x^3 + x + 2 \text{ in } \mathbb{Z}_5[x] \tag{1.20}$$

and

$$(x^2 - 2x - 1) * (x^3 + 2x - 2) = x^5 - 2x^4 + x^3$$
$$+ x^2 + 2x + 2 \text{ in } \mathbb{Z}_7[x]. \tag{1.21}$$

Again, in $\mathbb{Q}[x]$,

$$(2x^3 - \tfrac{3}{4}x + 1)(\tfrac{5}{7}x^2 - \tfrac{1}{5}x + \tfrac{1}{4}) = \tfrac{10}{7}x^5 - \tfrac{2}{5}x^4 - \tfrac{1}{28}x^3 + \tfrac{121}{140}x^2 - \tfrac{31}{80}x + \tfrac{1}{4}$$

which implies

$$(2x^3 + 1)(2x^2 + x + 1) = x^5 - x^4 - x^3 - x^2$$
$$+ x + 1 \text{ in } \mathbb{Z}_3[x] \tag{1.22}$$

but this time there is no corresponding statement in $\mathbb{Z}_5[x]$, because some of the rational coefficients do not have counterparts modulo 5. In general, if you are working in $\mathbb{Q}[x]$ and the 'h' register holds the sum, difference or product of the polynomials 'f' and 'g' then, if you change to a domain $\mathbb{Z}_p[x]$ in which all the polynomials have counterparts, the 'h' register will still hold the sum, difference or product respectively of the polynomials f and g. However, if p and q are different primes, you should be very careful about drawing similar conclusions when moving from $\mathbb{Z}_p[x]$ to $\mathbb{Z}_q[x]$, or even when changing from $\mathbb{Z}_p[x]$ back to $\mathbb{Q}[x]$. Try for instance multiplying $x - 3$ and $x + 3$ in $\mathbb{Z}_7[x]$. You should get $x^2 - 2$. If you now change to polynomial arithmetic modulo 13, (by pressing 'Ctrl-P' and then entering 13), the polynomials $x - 3$, $x + 3$ and $x^2 - 2$ will stay the same since coefficients between -6 and $+6$ are allowed here. However $x^2 - 2$ is *no longer the product* of $x - 3$ and $x + 3$. (What is their product?) So the same arithmetic operations may well produce different results if the underlying domains differ.

Activity 1.6 Just one of the equations displayed in (1.19)–(1.22) is also true in $\mathbb{Q}[x]$. Which one is it?

⟩1.3 Polynomial division

We have seen that subtraction can be defined in terms of addition. Division bears an entirely similar relation to multiplication, since dividing f by g means finding a polynomial q such that $f/g = q$, and that in turn means finding q so that $f = q * g$. For instance in $\mathbb{Q}[x]$

$$2x^5 + x^4 - 17x^3 + 8x^2 + 9x - 3 = (2x^3 - 5x^2 + 3) * (x^2 + 3x - 1)$$

which is the same as saying

$$\frac{2x^5 + x^4 - 17x^3 + 8x^2 + 9x - 3}{(x^2 + 3x - 1)} = 2x^3 - 5x^2 + 3.$$

When $f = q * g$, with q and g non-zero, we always have

$$\deg(f) = \deg(q) + \deg(g) \geqslant \deg(g) \tag{1.23}$$

where $\deg(f)$ is short for 'the degree of f' (and likewise for $\deg(g)$, $\deg(q)$). So we should only expect g to divide into the polynomial f if

$\deg(f) \geqslant \deg(g)$. Even then being able to divide one polynomial into another is the exception rather than the rule.

Example 1.4
The polynomial x does not divide into the polynomial $x^2 + 2x - 3$ (unless we are working in $\mathbb{Z}_3[x]$). This is because $x^2 + 2x - 3$ cannot be expressed as $x * q$ for any polynomial q. Since x and $x^2 + 2x - 3$ have degrees 1 and 2 respectively, (1.23) shows that such a quotient q would have to have degree 1 and so be of the form $ax + b$ for some numbers a and b. But $x * (ax + b) = ax^2 + bx$ which lacks the term '-3'. In this case the nearest that we can get to division is to write $x^2 + 2x - 3 = x * (x + 2) - 3$.

In the general situation where g does not divide f we can still express f as a multiple of g plus a remainder of smaller degree. That is, we can find polynomials q and r with

$$f = q * g + r \text{ and } \deg(r) < \deg(g). \tag{1.24}$$

The easiest way to see this is by using induction on the degree of f. The result is certainly true if $\deg(f) < \deg(g)$, for then we can write $f = 0 * g + f$ with f itself as the remainder of smaller degree than the divisor. If $\deg(f) \geqslant \deg(g)$, with f not divisible by g, we write

$$f(x) = a_k x^k + a_{k-1} x^{k-1} + \ldots + a_0$$

and

$$g(x) = b_n x^n + b_{n-1} x^{n-1} + \ldots + b_0$$

where $k \geqslant n$. We may assume inductively that any polynomial of degree less than k which is not divisible by g can be written as a multiple of g plus a remainder of smaller degree. Then

$$f - \frac{a_k}{b_n} x^{k-n} g = (a_{k-1} - \frac{a_k}{b_n} b_{n-1}) x^{k-1} + \ldots + (a_{k-n} - \frac{a_k}{b_n} b_0) x^{k-n} + \ldots + a_0.$$

So $f - \frac{a_k}{b_n} x^{k-n} g$ is a polynomial of degree $k - 1$ or less and is not divisible by g (since f is not). Therefore, by assumption, it can be written as $q_1 * g + r$ for some q_1 and r with $\deg(r) < \deg(g)$. This means that

$$f = \left(\frac{a_k}{b_n} x^{k-n} + q_1 \right) * g + r$$

which is the desired type of decomposition for f. The assertion now follows for any polynomial.

The case when g divides exactly into f is the case when the remainder r in (1.24) is the zero polynomial. So for *any* two polynomials f and g we can always find q and r, in the same domain as f and g, that satisfy

$$f = q * g + r \; with \; either \; r = 0 \; or \; \deg(r) < \deg(g). \qquad (1.25)$$

Note that in (1.25) there is only one possible choice of quotient and remainder that makes the remainder either zero or of smaller degree than g. For suppose that we have expressed f as $q * g + r$ and also as $q_1 * g + r_1$ with both r and r_1 either zero or of smaller degree than g. Then $q * g + r = q_1 * g + r_1$ implies

$$(q - q_1) * g = r_1 - r.$$

If $q \neq q_1$ the left side of this last equation would be a non-zero multiple of g and so its degree would be at least $\deg(g)$ which is larger than the degree of $r_1 - r$. So necessarily we have $q = q_1$ and thus $r = r_1$. The key marked '/' will instruct the computer to perform division with remainder. In $\mathbb{Q}[x]$ for instance, if we ensure that the 'f' register contains $\frac{2}{3}x^4 - 7x^3 + \frac{3}{4}x^2 + 1$, and the '$g$' register contains $x^2 - 3x + 2$, then entering 'f/g' produces a quotient of $\frac{2}{3}x^2 - 5x - \frac{187}{12}$ and a remainder of $-\frac{147}{4}x + \frac{193}{6}$. The computer always calls the quotient of any division operation 'q', and the remainder 'r'. Of course we can also ask the computer to perform other divisions such as 'g/h', or even 'q/r', though in the latter case the original contents of the 'q' and 'r' registers would be overwritten with the new quotient and remainder. When working in $\mathbb{Q}[x]$ the computer sometimes encounters exceptionally large numbers which it cannot deal with easily. This is usually unavoidable and, where division is concerned, is often unexpected. For instance if we try to divide x^9 by $x - 10$ the computer starts to meet numbers that are too large for it, even though the initial coefficients are small. Here this is because
$x^9 = (x - 10)$
$*(x^8 + 10x^7 + 100x^6 + 1000x^5 + 10\,000x^4 + 100\,000x^3 + 1\,000\,000x^2 + 10\,000\,000x + 100\,000\,000) + 1\,000\,000\,000.$
So several of the coefficients in both quotient and remainder exceed the computer's working limit of 32768. Before doing any arithmetic operation the computer always checks to see whether any numbers are going to overflow its internal limits. If so, it issues a warning and will not attempt the operation.

Activity 1.7 In each of the following cases express $f(x)$ as $q(x) *$ $g(x) + r(x)$ where $r(x)$ is either 0 or has smaller degree than $g(x)$.
(i) $f(x) = 25x^4 - 7x^3 + 19x^2 + 1$ and $g(x) = x^2 - 3x + 2$ in $\mathbb{Z}_{73}[x]$;
(ii) $f(x) = 25x^2 - 5x + 27$ and $g(x) = x^2 - 3x + 2$ in $\mathbb{Z}_{73}[x]$;
(iii) $f(x) = 12x^4 - 7x^3 + 6x^2 + 1$ and $g(x) = x + 5$ in $\mathbb{Z}_{29}[x]$;
(iv) $f(x) = 2x^7 - 5x^5 + 3x^4 + 12x^3 - 5x^2 - 29x + 6$ and $g(x) = 2x^3 - 5x + 1$ in $\mathbb{Q}[x]$;
(v) $f(x) = 2x^4 + 2x^3 - x + 1$ and $g(x) = x^2 + 2$, first in $\mathbb{Q}[x]$, then in $\mathbb{Z}_5[x]$ and $\mathbb{Z}_2[x]$.

⟩1.4 Common divisors of polynomials

Although it can be quite difficult to find all the polynomials that exactly divide into a given one, it is easy to apply division with remainder repeatedly to see if *two* given polynomials have any common polynomial divisors. The essential idea is just the same as that of Euclid's algorithm for finding the greatest common divisor, or GCD, of two ordinary integers. A GCD of two polynomials is a common divisor which is divisible by any other common divisor and the method is best illustrated by using it to find a GCD of two specific polynomials, say $x^5 - 2x^4 + x - 2$ and $x^3 - x^2 - x - 2$ in $\mathbb{Q}[x]$. We first put $x^5 - 2x^4 + x - 2$ into the 'f' register and $x^3 - x^2 - x - 2$ into the 'g' register (it doesn't actually matter which is called 'f' and which 'g'). We then divide f by g (by typing 'f/g'), which shows that

$$f = (x^2 - x) * g + (x^2 - x - 2). \tag{1.26}$$

This equation shows that any common divisor of f and g would have to divide the remainder $x^2 - x - 2$ as well; and any common divisor of g and $x^2 - x - 2$ would also have to divide f. So $\text{GCD}(f, g) = \text{GCD}(g, x^2 - x - 2)$. We continue by moving these polynomials, g and $x^2 - x - 2$, into the 'f' and 'g' registers by typing '$f = g$' and then '$g = r$'. Typing 'f/g' now produces

$$x^3 - x^2 - x - 2 = x * (x^2 - x - 2) + (x - 2) \tag{1.27}$$

which shows as above that $\text{GCD}(x^3 - x^2 - x - 2, x^2 - x - 2) = \text{GCD}(x^2 - x - 2, x - 2)$. Again repeat the steps '$f = g$', '$g = r$' and then 'f/g' to obtain

$$x^2 - x - 2 = (x + 1) * (x - 2). \tag{1.28}$$

This last equation says that $x - 2$ is itself a divisor of $x^2 - x - 2$, which means that $\mathrm{GCD}(x^2 - x - 2, x - 2)$ is $x - 2$, and that must also be a GCD of the original polynomials f and g.

The equations (1.26)–(1.28) are an example of *Euclid's algorithm for polynomials*. The general algorithm can be applied similarly to find a GCD of any two polynomials and for convenience of use on the computer it can be stated as follows.

First put the polynomials (in either order) into the 'g' and 'r' registers. Then keep entering

$$f = g$$
$$g = r$$
$$f/g$$

until $r = 0$.

A GCD of the original pair of polynomials is then given by 'g'.

The algorithm always terminates because the degrees of the remainders are strictly decreasing.

Before we try some more examples there is one point to be cleared up. I am sure you will have noticed that, in the above discussion, I kept referring to 'a GCD' rather than '*the* GCD'. This is because there are actually several possible GCDs of any given pair of polynomials. The polynomial $x - 2$ is, as we have seen, a GCD of $x^5 - 2x^4 + x - 2$ and $x^3 - x^2 - x - 2$; but so also is $\frac{2}{3} * (x - 2) = \frac{2}{3}x - \frac{4}{3}$. Certainly it is a common divisor because $x^5 - 2x^4 + x - 2 = (\frac{3}{2}x^4 + \frac{3}{2}) * (\frac{2}{3}x - \frac{4}{3})$ and $x^3 - x^2 - x - 2 = (\frac{3}{2}x^2 + \frac{3}{2}x + \frac{3}{2}) * (\frac{2}{3}x - \frac{4}{3})$; and it is a greatest common divisor since it is divisible by the greatest common divisor $x - 2$. Similarly if h is a GCD of any two polynomials f and g and a is a number then ah is also a GCD, and likewise any GCD is a constant multiple of h (see section 2 in Appendix 1).

Activity 1.8 Use Euclid's algorithm to find a GCD for each of the following pairs of polynomials.
(i) $x^6 - 5x^5 + 6x^4 + x^2 - 5x + 6$ and $2x^4 - 9x^3 + 8x^2 + x + 6$ in $\mathbb{Q}[x]$, $\mathbb{Z}_{29}[x]$ and then in $\mathbb{Z}_2[x]$;
(ii) $x^4 + 1$ and $x^3 - x^2 - x - 2$ in $\mathbb{Z}_{13}[x]$, $\mathbb{Z}_{17}[x]$ and in $\mathbb{Q}[x]$;
(iii) $x^3 + x^2 - x - 1$ and $x^3 + 1$ in $\mathbb{Z}_3[x]$ and in $\mathbb{Q}[x]$.

Notice what happens if we apply the method to find the GCD of the numbers 6 and 10 in $\mathbb{Q}[x]$. The algorithm halts after one step since

10 divides 6 ($6 = \frac{3}{5} * 10$). This is because we are allowing rational coefficients which means that any non-zero number divides any other. In particular any rational number is a GCD for 6 and 10 in $\mathbb{Q}[x]$.

The computer will actually work out a GCD entirely automatically for you if you enter 'f, g', 'g, q' or any similar expression. It always puts the result in the 'h' register and ensures that the leading coefficient of the displayed GCD is one.

Activity 1.9 Get the computer to check your answers to Activity 1.8. Remember that two GCDs will be equivalent if they differ by a constant multiple (which may be checked by asking the computer to divide one into the other).

In the domains $\mathbb{Q}[x]$ and $\mathbb{Z}_p[x]$ an important property of a GCD is that any greatest common divisor of two polynomials can be expressed as a linear combination of them. That is, if h is a GCD of f and g, then there are polynomials $u(x)$ and $v(x)$, in the same domain as f and g, such that

$$h(x) = u(x) * f(x) + v(x) * g(x). \tag{1.29}$$

This is just like the situation in $\mathbb{Z}$ where, if a and b are integers with GCD d, there are integers k and l such that $d = ka + lb$. We can also prove the result for polynomials with similar reasoning to that for integers. We assume that f and g are in some fixed domain (either $\mathbb{Q}[x]$ or a $\mathbb{Z}_p[x]$) and consider the collection S of all linear combinations $uf + vg$ formed as $u(x)$ and $v(x)$ each range over all the polynomials in the domain. At least one of these linear combinations will have smallest degree out of all the non-zero polynomials in S. If $h_1(x)$ is such a polynomial of least degree in S then

$$h_1(x) = u_1(x) * f(x) + v_1(x) * g(x) \tag{1.30}$$

for some specific u_1 and v_1. This h_1 will turn out to be a divisor of f, as we can see by using division with remainder, for there will be polynomials q and r such that

$$f = q * h_1 + r$$

where r is either zero or $\deg(r) < \deg(h_1)$. Therefore $r = f - q * (u_1 f + v_1 g) = (1 - qu_1) * f + (-qv_1) * g$ which is one of the combinations in S. There is no element of S that has smaller degree than h_1; so we cannot have $\deg(r) < \deg(h_1)$ and are left with $r = 0$, which means that h_1 divides f. Precisely similar reasoning shows that h_1 divides g, so that it is a common divisor of f and g. Furthermore, any other common divisor of f and g will divide the right side of (1.30) and so divide h_1. Thus h_1 is definitely a GCD of f and g. This is not yet exactly what we wanted, since we started with a given GCD h of f and g, and have proved that some other GCD h_1 is a linear combination of the desired form. However any GCD is a constant multiple of any other, so h will be αh_1 for some number α, and this means that $h = \alpha u_1 f + \alpha v_1 g$ as required. This establishes (1.29).

For any polynomials f and g with a GCD h, the computer will try to find suitable multipliers u and v that satisfy (1.29) if you enter '$f, g\$$'. It will actually put these multipliers in the 'q' and 'r' registers but, to avoid any confusion, the menu option will remind you of the names used. The computer should never run into trouble when it is working in a domain $\mathbb{Z}_p[x]$ for a prime p, but it may encounter overflow problems in $\mathbb{Q}[x]$. It will always tell you if it does not think it can manage all the arithmetic involved.

Activity 1.10 For each domain and pair of polynomials in Activity 1.8, use the computer to express a GCD of the polynomials as a linear combination of them, with multipliers from the given domain.

〉1.5 Units, irreducibles and the factor theorem

It is usually more difficult to find all the divisors, or factors, of a single polynomial; though some can be noted immediately since they occur as divisors of every polynomial. These are the *unit* or *invertible* polynomials and, as the latter name suggests, they are the polynomials that have polynomial reciprocals. So $u(x)$ is a unit when there is a polynomial $u_1(x)$ satisfying

$$u * u_1 = 1. \tag{1.31}$$

If so, then u is a divisor of every polynomial, since for any f we can write $f = u * (u_1 f)$ with the quotient $u_1 f$ being another polynomial. Note that (1.31) gives $\deg(u) + \deg(u_1) = \deg(1) = 0$, and so implies $\deg(u) = 0$. Thus

$$\textit{unit polynomials have to be constants,} \tag{1.32}$$

and in $\mathbb{Q}[x]$ or any $\mathbb{Z}_p[x]$ all non-zero constants are units. [In section 1.7 we shall see an example of a polynomial domain in which there are some constants that are not units.] Any polynomial f will also have another obvious class of divisors since it must be divisible by all unit multiples of itself. If u is a unit with reciprocal u_1, then uf divides f because $f = u_1 * (uf)$. A polynomial, that is not zero and not a unit, is called *irreducible* or *prime* if it only has those two types of divisors (units and unit multiples of itself). For example the polynomial x is divisible by every constant α and every multiple αx. As we shall see in Example 1.5 it has no other divisors and so is irreducible. On the other hand, the polynomial $x^2 + x$ has factors such as x and $x + 1$ as well as the obvious kinds. So it is neither a unit nor irreducible and is called *composite* . A non-zero polynomial f, which is not a unit and not irreducible,† (in other words is composite), must be divisible by some factor g that is neither a unit nor a unit multiple of f and we can thus write

$$f = g * q \tag{1.33}$$

for some g and q. We then have

$$\deg(f) = \deg(g) + \deg(q) \geqslant 2 * \min(\deg(g), \deg(q)). \tag{1.34}$$

This means that if f does have some factors, other than the trivial ones that we have mentioned above, it must have a factor with degree at most $\frac{1}{2}\deg(f)$.

Example 1.5
In $\mathbb{Q}[x]$ or $\mathbb{Z}_p[x]$ every *linear*‡ polynomial is irreducible. For if f was a linear polynomial that had the typical factorization (1.33), the inequality (1.34) would imply that one of the factors would be of degree 0 and so be a constant, say a. Then the other factor would be $\frac{1}{a} * f$. In other words f could not have any factors other than units or unit multiples of f; which would mean that it was irreducible.

Example 1.6
There are only two different linear polynomials in $\mathbb{Z}_2[x]$. This is because, in $\mathbb{Z}_2[x]$, the only possible coefficients are 0 and 1. So in a linear

---

† The opposite of the word 'irreducible' is of course 'reducible', but for some reason 'reducible' is not as frequently used in mathematics as 'composite'.
‡ As usual, the word 'linear' refers to a polynomial of degree 1.

polynomial the non-zero coefficient of x must be 1 and the polynomial is either x or $x + 1$. Similarly the only *quadratic* polynomials (those of degree 2) in $\mathbb{Z}_2[x]$ are x^2, $x^2 + x$, $x^2 + 1$ and $x^2 + x + 1$.

If we have a polynomial of degree 2 or 3, the remark after (1.34) means that we only have to see if it has factors of degree 1 in order to know whether it is irreducible or not. For a polynomial of degree 4 or 5 we would have to be sure that there were no linear or quadratic factors before we could say that the polynomial was irreducible.

There is a simple observation which helps in testing for divisibility by a linear polynomial. Suppose we want to see if $ax + b$ is a factor of $f(x)$. We use division with remainder, as in (1.25), but here the remainder, being of smaller degree than the divisor $ax + b$, will be of degree zero. So we will have an equation like

$$f(x) = q(x) * (ax + b) + r \qquad (1.35)$$

where r in this case is a number. We can find its value if we substitute $x = -b/a$ in (1.35), for this makes the linear term $ax + b$ equal to zero and gives $f(-b/a) = r$. So without knowing the quotient at all, we can find the remainder on dividing f by $ax + b$ by evaluating $f(-b/a)$. Of course $ax + b$ will be a factor of f precisely when the remainder is zero, which is to say that

$ax + b$ *is a factor of* $f(x)$ *if and only if* $f(-b/a) = 0$.   (1.36)

This important little result is often known as *the factor theorem*. A frequently encountered special case is when $a = 1$, and (1.36) then becomes

$x + b$ *is a factor of* $f(x)$ *if and only if* $f(-b) = 0$.   (1.37)

A number α such that $f(\alpha) = 0$ is a *root* of f so, with $\alpha = -b$, (1.37) can be restated as

α *is a root of* f *if and only if* $x - \alpha$ *is a factor of* f.   (1.38)

Example 1.7
(i) In $\mathbb{Q}[x]$ the numbers 2 and $\frac{5}{3}$ are roots of $f(x) = 3x^2 - 11x + 10$ since $f(2) = 0$ and $f(\frac{5}{3}) = 0$. So $x - 2$, $x - \frac{5}{3}$, and also $3(x - \frac{5}{3}) = 3x - 5$, are factors of f. Indeed $f(x) = (x - 2)(3x - 5) = 3(x - 2)(x - \frac{5}{3})$.

(ii) In $\mathbb{Q}[x]$, $2x - 1$ does not divide the polynomial $f(x) = 2x^4 + 3x^3 - x^2 - x$ since $f(\frac{1}{2}) \neq 0$. However $2x + 1$ is a factor since $f(-\frac{1}{2}) = 0$ and $x = x - 0$ is also a factor since 0 is a root. The polynomial factorizes as $x(2x + 1)(x^2 + x - 1)$.

(iii) In $\mathbb{Z}_{11}[x]$, the numbers 2, 3 and -3 are roots of $x^3 - 2x^2 + 2x - 4$ and the polynomial is $(x - 2)(x - 3)(x + 3)$.

Activity 1.11 In this activity you should calculate—at least some of—the remainders by hand and then, if appropriate, use the computer to check your answers.

(i) In $\mathbb{Q}[x]$ what is the remainder when $9x^5 - x + 2$ is divided by $3x - 1$?; when $2x^4 - 9x^3 - \frac{22}{3}x^2 + \frac{38}{3}x - 5$ is divided by $2x + 3$?; when $x^3 - 4x^2 + 1$ is divided by $x - 5$?

(ii) What is the remainder when $x^4 + x^3 + x^2 + x + 1$ is divided by $x - 1$ in $\mathbb{Z}_5[x]$?; when $x^{p-1} + x^{p-2} + \ldots + x + 1$ is divided by $x - 1$ in $\mathbb{Z}_p[x]$?; when $x^3 - 4x^2 + 1$ is divided by $x - 5$ in $\mathbb{Z}_{13}[x]$?

(iii) If $f(x) = x^3 - 2x^2 - 5x - 1$ in $\mathbb{Z}_{11}[x]$, calculate $f(0)$, $f(1), \ldots, f(10)$. So determine the roots of f in $\mathbb{Z}_{11}$ and factorize f.

Activity 1.12 Only one of the four quadratic polynomials in $\mathbb{Z}_2[x]$ is irreducible. Which one is it? (Test for divisibility by each of the linear polynomials to see which quadratic polynomial does not have factors of degree 1.)

$\rangle$1.6 Factorization into irreducible polynomials

Any composite polynomial can be expressed as the product of—perhaps several—irreducible factors of smaller degree. In $\mathbb{Q}[x]$ the polynomial $x^7 + 3x^6 + 4x^5 + 4x^4 - x^3 - 4x^2 - 5x - 2$, for instance, is composite and can be written as $(x^3 + 3x^2 + 3x + 2) * (x^4 + x^2 - x - 1)$. Neither of these factors is prime, however, and it can actually be written as $(x + 2) * (x^2 + x + 1) * (x - 1) * (x^3 + x^2 + 2x + 1)$ which is a product of prime factors. In general this is easy to see using induction on the degree of the polynomial we start with. There are no composite polynomials of degree 1 and, by (1.34), any composite quadratic polynomial can only be a product of two linear polynomials, which of course will each be irreducible. So we may assume inductively that the result has been proved for all composite polynomials of degree less than some number $n > 2$. Then if f is a composite polynomial of degree n it will be a product of two polynomials g and h of smaller degree: $f = g * h$. This makes $\deg(g) < n$ and $\deg(h) < n$ and so, by assumption, g and h will each be either irreducible or a product of irreducible factors. If say $g = p_1 * p_2 * \ldots * p_r$ and $h = q_1 * q_2 * \ldots * q_s$ where $p_1, \ldots, p_r$, $q_1, \ldots, q_s$ are prime polynomials and $r \geqslant 1$, $s \geqslant 1$, then putting these expressions together we have $f = p_1 * p_2 * \ldots * p_r * q_1 * q_2 * \ldots * q_s$ which completes the inductive step.

If f is a composite polynomial in any of the domains we have met so far ($\mathbb{Q}[x]$, $\mathbb{Z}_p[x]$), it is not only true that f can be expressed as a product of prime polynomials but that there is essentially only one way to express it as such a product. Here, 'essentially' means that the order of the factors can be changed, or some of them can be multiplied by constants provided that others are divided by compensating constants, but those are the only changes we can make and still have the same product. For example, in $\mathbb{Q}[x]$, the prime factorization of $2x^6 - 3x^5 + 2x^4 + 13x^3 - 21x^2 + 14x - 7$ can be written as

$$(x - 1)(2x^2 - x + 1)(x^3 + 7)$$

or

$$(x^3 + 7)(2x^2 - x + 1)(x - 1)$$

or

$$(2x^2 - x + 1)(-x^3 - 7)(-x + 1)$$

or even as

$$(4x^2 - 2x + 2)(-\tfrac{1}{6}x^3 - \tfrac{7}{6})(-3x + 3)$$

but we cannot write the same polynomial as a product involving a different number of prime factors, or with any totally different prime polynomial appearing as one of the factors. This will follow if we show that any prime polynomial dividing $(x - 1)(2x^2 - x + 1)(x^3 + 7)$ must divide either $x - 1$ or $2x^2 - x + 1$ or $x^3 + 7$, for that will force it to be a constant multiple of one of those prime factors.

We shall show first that if a polynomial $p(x)$ is irreducible and

> *if p divides a product fg of two factors then it must*   (1.39)
> *divide one of the terms f, g.*

We may suppose that f is not divisible by p, and so not by any unit multiple of p, which means that the only common divisors of p and f

are units (as units and unit multiples of p are the only possible divisors of p). Thus each unit is a GCD of p and f. In particular, 1 is a GCD and so there will be polynomials $a(x)$ and $b(x)$ such that

$$a(x)p(x) + b(x)f(x) = 1$$

and therefore

$$a(x)p(x)g(x) + b(x)f(x)g(x) = g(x).$$

Here p divides apg and also bfg since it divides fg. So it must divide g on the right-hand-side and that gives (1.39). Appendix 1 discusses the counterparts of irreducibility and of property (1.39) in general integral domains. In particular it is shown in section A1.3 that in any domain an element p that satisfies (1.39) must be irreducible. There are integral domains where not all the irreducible elements satisfy (1.39); and in those domains the word 'prime' is only applied to the irreducible elements that do have that property. In Chapter 4 we shall meet an example of such an integral domain in which there are some irreducible elements that do not satisfy (1.39).

Property (1.39) can easily be extended to the case where an irreducible polynomial p divides a product $f_1 f_2 \ldots f_n$ of several polynomials. For if p divides $f_1 * f_2 f_3 \ldots f_n$, it must divide either f_1 or $f_2 * f_3 \ldots f_n$. Then it must divide either f_1, f_2 or $f_3 \ldots f_n$, and ultimately either f_1, f_2, ..., or f_n. As a special case we get the result, illustrated above, that $x - 1$, $2x^2 - x + 1$, $x^3 + 7$, or constant multiples of them, are the only possible irreducible factors which divide $2x^6 - 3x^5 + 2x^4 + 13x^3 - 21x^2 + 14x - 7$.

We can also deduce the general result that, in either $\mathbb{Q}[x]$ or $\mathbb{Z}_p[x]$, any polynomial of degree 1 or more is either irreducible or can be represented in essentially only one way as a product of irreducibles. That is to say, a polynomial that is neither zero nor a unit must either be irreducible or, if it can be expressed as two products $p_1 p_2 \ldots p_r$ and $q_1 q_2 \ldots q_s$ of irreducible factors, then $r = s$ and each q_j is a unit multiple of a p_i. This result is true for linear polynomials since they are all irreducible. Suppose we have proved it for all polynomials of degree less than some $n \geqslant 2$, and that f is a polynomial of degree n which can be represented as $p_1 p_2 \ldots p_r$ and also as $q_1 q_2 \ldots q_s$ for irreducible polynomials $p_1, \ldots, p_r, q_1, \ldots, q_s$. Then, from $p_1 p_2 \ldots p_r = q_1 q_2 \ldots q_s$, we see that p_1 must divide one of $q_1, \ldots, q_s$. Say $q_1 = \alpha p_1$. But the only way this can hold, with p_1 and q_1 irreducible, is for α to be a unit. So $p_2 \ldots p_r = \alpha q_2 \ldots q_s$ and this last product, being $\frac{f}{p_1}$, is a polynomial of smaller degree than n. The assumption that the result is true for all polynomials of degree less than n now implies that $r = s$ and αq_2 (and so q_2), ..., q_r are each constant multiples of $p_2, ..., p_r$ (though perhaps in a different order). This gives the result.

〉1.7 Polynomials with integer coefficients

The set of all polynomials with unrestricted integer coefficients is called $\mathbb{Z}[x]$. It is contained in $\mathbb{Q}[x]$ and is again an integral domain. Units and irreducibles in $\mathbb{Z}[x]$ are defined just as in section 1.5 and again units have to be constants. The constants are ordinary integers, but only two of them are units here, since only 1 and -1 have inverses that are also in $\mathbb{Z}[x]$. All the other non-zero integers have the same classifications of prime or composite in $\mathbb{Z}[x]$ as they do in $\mathbb{Z}$. There is also a unique factorization property, since each composite polynomial in $\mathbb{Z}[x]$ can be written in essentially just one way as a product of irreducibles in $\mathbb{Z}[x]$ (see Exercises 7–10 below). However the resulting representation might not resemble its representation as a product of irreducibles in $\mathbb{Q}[x]$. In particular polynomials can be irreducible in $\mathbb{Q}[x]$ but composite in $\mathbb{Z}[x]$. For instance $10x + 10$ is irreducible in $\mathbb{Q}[x]$ but is composite in $\mathbb{Z}[x]$, being $2 * 5 * (x + 1)$. However

> *if a non-constant polynomial with integer*   (1.40)
> *coefficients is irreducible in $\mathbb{Z}[x]$*
> *then it must also be irreducible in $\mathbb{Q}[x]$.*

The first step towards seeing this is to note that any polynomial with rational coefficients is a multiple of one with integer coefficients. For example $\frac{2}{3}x^3 - \frac{4}{15}x^2 + 6x - \frac{10}{21}$ can be written as $\frac{2}{105}(35x^3 - 14x^2 + 315x - 25)$. A rational polynomial of degree n say must similarly be of the form

$$\frac{a_n}{b_n}x^n + ... + \frac{a_0}{b_0} \tag{1.41}$$

where $a_0, ..., a_n, b_0, ..., b_n$ are integers. If b is the least common multiple of $b_0, ..., b_n$, we can express the polynomial as $\frac{1}{b}(k_n x^n + ... + k_0)$ for some integers $k_0, ..., k_n$; and then if a is the GCD of $k_0, ..., k_n$, we can write that as

$$\frac{a}{b}(c_n x^n + ... + c_0) \tag{1.42}$$

where $c_0 = \frac{k_0}{a}, \ldots, c_n = \frac{k_n}{a}$ are integers with no common factor bigger than 1. A non-zero polynomial, like $c_n x^n + \ldots + c_0$, whose coefficients are integers with GCD 1, is called *primitive* and, no matter how we do it, there is actually only one way to write a polynomial in $\mathbb{Q}[x]$ as the product of a positive rational number and a primitive polynomial. For suppose we were able to express some polynomial as $\frac{a}{b} * f$ and also as $\frac{c}{d} * g$ with a, b, c, d positive integers and f and g both primitive. Then $\frac{a}{b} * f = \frac{c}{d} * g$ would imply $ad * f = bc * g$. But $ad * f$ would be a polynomial in $\mathbb{Z}[x]$ with ad being the GCD of its coefficients; and similary bc would be the GCD of the coefficients of $bc * g$ ($= ad * f$). Therefore $ad = bc$, implying $\frac{a}{b} = \frac{c}{d}$, and then $f = g$. The unique positive rational multiplier $\frac{a}{b}$ is called the *rational content* of the polynomial.

Example 1.8
The rational content of a polynomial f in $\mathbb{Z}[x]$ must be an integer. This is because f can certainly be written as $a * g$ where the integer a is the GCD of the coefficients of f and g is primitive; and by the last paragraph that is the only way to write f with a positive rational multiplier.

Example 1.9
If a polynomial f in $\mathbb{Z}[x]$ is a rational multiple of another polynomial g in $\mathbb{Z}[x]$, and if f and g are both primitive, then $f = g$ or $f = -g$. This is because $1 * f$ must be the only way to write f as a positive rational multiple of a primitive polynomial. So if $f = a * g$ with a rational, then $f = |a| * (\pm g)$, so that $|a| = 1$ and $a = \pm 1$.

Example 1.10
If f is a polynomial in $\mathbb{Z}[x]$ which is irreducible, and if $\deg(f) \geqslant 1$, then f is primitive. For otherwise we could write f as $kg(x)$ with $k > 1$ and g primitive. But then neither k nor g would be units in $\mathbb{Z}[x]$ and so f would be composite.

We can now see why a non-constant polynomial that is irreducible in $\mathbb{Z}[x]$ must also be irreducible in $\mathbb{Q}[x]$. It is actually easier to show this the other way round: that if a polynomial with integer coefficients has a factorization with both factors in $\mathbb{Q}[x]$ then it must have one where both factors are in $\mathbb{Z}[x]$ and have the same degrees as the respective factors in $\mathbb{Q}[x]$. Suppose then that

$$f = g * h$$

where f is in $\mathbb{Z}[x]$ and g and h are both in $\mathbb{Q}[x]$. Say $g = \alpha * g_1$, $h = \beta * h_1$ where the positive numbers α and β are rational and g_1, h_1 are primitive. Then

$$f = \alpha\beta * (g_1 h_1)$$

and, because the product $g_1 h_1$ is primitive (see Appendix 2), this must be the unique expression of f as a positive rational times a primitive polynomial. So, from Example 1.8, $\alpha\beta$ is an integer, and $f = (\alpha\beta g_1) * h_1$ is then a factorization with both factors in $\mathbb{Z}[x]$.

〉1.8 Factorization in $\mathbb{Z}_p[x]$ and applications to $\mathbb{Z}[x]$

In an integral domain of the form $\mathbb{Z}_p[x]$ there are only p possible coefficients and so only a finite number of polynomials of any given degree. There are p constant polynomials (the p different numbers in $\mathbb{Z}_p$); $p(p-1)$ polynomials of degree 1 (since there are p choices for the coefficient of x^0 and $p-1$ choices for the *non-zero* coefficient of x); $p^2(p-1)$ of degree 2, ... So, in principle, it is possible to list all the polynomials of small degree and thus test whether a given polynomial of degree n has factors by dividing it by each polynomial whose degree is between 1 and $\frac{n}{2}$. We can also use the fact that every factor is a multiple of one with leading coefficient 1. So, in searching for linear or quadratic factors for instance, it is only necessary to examine the p linear possibilities of the form $x + b$ where $b \in \mathbb{Z}_p$ and the p^2 quadratic ones of the form $x^2 + bx + c$, where b and c are in $\mathbb{Z}_p$. Of course for even moderately large values of p and n there are still too many polynomials of small degree for this procedure to be really practicable, but it is quite workable when p and n are small.

Activity 1.13 Factorize each of the polynomials $x^3 - 1$, $x^3 - 2$, $x^3 - 3$, $x^3 - 4$ into irreducible factors in $\mathbb{Z}_5[x]$. (You will find an easy way to check your answers later on.)

In section 1.2 we saw that each factorization in $\mathbb{Q}[x]$ or $\mathbb{Z}[x]$ gives rise to a factorization in $\mathbb{Z}_p[x]$. The other side of this statement is that if a polynomial is irreducible in $\mathbb{Z}_p[x]$ for a certain p, then, when the coefficients are interpreted as integers rather than as elements of $\mathbb{Z}_p$, the polynomial is irreducible in $\mathbb{Z}[x]$ (and so in $\mathbb{Q}[x]$). For instance, by testing as possible divisors the five linear polynomials $x, x \pm 1, x \pm 2$

in $\mathbb{Z}_5[x]$, it is easy to check that $x^3 - x - 2$ is irreducible in $\mathbb{Z}_5[x]$ and so it is immediate that $x^3 - x - 2$ is irreducible in $\mathbb{Z}[x]$ and $\mathbb{Q}[x]$. It also follows immediately that $6x^3 - x + 3$, for example, is irreducible in $\mathbb{Z}[x]$ since it is equivalent to the same polynomial $x^3 - x - 2$ in $\mathbb{Z}_5[x]$. Note how we are led astray though if we try to use the same reasoning with the polynomial $15x^3 - 4x^2 + x + 2$ in $\mathbb{Z}[x]$. This is equivalent to $x^2 + x + 2$ in $\mathbb{Z}_5[x]$ which is again irreducible there. Thus any factorization of $15x^3 - 4x^2 + x + 2$ in $\mathbb{Z}[x]$ must reduce modulo 5 to a factorization of $x^2 + x + 2$ in $\mathbb{Z}_5[x]$ and so to a constant times a quadratic; but that is not enough to conclude that $15x^3 - 4x^2 + x + 2$ is irreducible. In fact it is $(5x + 2)(3x^2 - 2x + 1)$. So if a polynomial f in $\mathbb{Z}[x]$ is equivalent to f_p in $\mathbb{Z}_p[x]$, then the irreducibility of f_p in $\mathbb{Z}_p[x]$ only enables us to deduce the irreducibility of f in $\mathbb{Z}[x]$ when f and f_p have the same degree or, in other words, when the leading coefficient of f is not divisible by p.

Activity 1.14 For each of the following polynomials, decide whether there is a prime p, less than 10 say, for which the polynomial is irreducible in $\mathbb{Z}_p[x]$. Are any of them irreducible in $\mathbb{Z}[x]$? (i)$2x^3 - 3x^2 - 3x + 1$; (ii)$3x^2 + 2x - 5$; (iii)$3x^5 + x^4 - x^3 + 5x - 3$; (iv)$6x^3 - x - 2$; (v)$3x^3 - 6x^2 - 2$; (vi)$14x^3 - 15x^2 + 25x - 12$.

Even if a polynomial is composite in some domain $\mathbb{Z}_p[x]$, the form of its factorization there can give us valuable insights into the nature of any factorization in $\mathbb{Z}[x]$. For example consider a polynomial f in $\mathbb{Z}[x]$ all of whose coefficients, except the leading one, are divisble by a prime p in $\mathbb{Z}$. So

$$f(x) = a_k x^k + a_{k-1} x^{k-1} + \ldots + a_0 \equiv a_k x^k (\bmod p).$$

Suppose f is composite in $\mathbb{Z}[x]$, and can be written as

$$f(x) = g(x)h(x)$$

where the factors g and h are in $\mathbb{Z}[x]$ and each have degree at least 1. When reduced modulo p, this decomposition gives

$$f_p(x) = g_p(x)h_p(x) \text{ in } \mathbb{Z}_p[x]$$

where $f_p(x) = a_k x^k$ and $\deg(g_p) = \deg(g) \geqslant 1$, $\deg(h_p) = \deg(h) \geqslant 1$. It is obvious that f_p is a product of k irreducible factors, each of which is a constant multiple of x. So, by the uniqueness of factorization in $\mathbb{Z}_p[x]$, g_p and h_p must be composed of (unit multiples of) these same linear factors. Thus $g_p(x) = bx^r$ and $h_p(x) = cx^s$, where $bc \equiv a_k$ (mod p). What this says about $g(x)$ and $h(x)$ in $\mathbb{Z}[x]$ is that $g(x) = b_r x^r + b_{r-1} x^{r-1} + \ldots + b_0$ and $h(x) = c_s x^s + c_{s-1} x^{s-1} + \ldots + c_0$ where $b_r \equiv b$ (mod p), $c_s \equiv c$ (mod p), and each of $b_{r-1}, \ldots, b_0, c_{s-1}, \ldots, c_0$ is divisible by p. So, when looking for factors of f in $\mathbb{Z}[x]$, we need only consider possible factors which again have all coefficients, except the leading one, divisible by p. More than that is true because, from the factorization $f(x) = g(x)h(x)$, we have $a_0 = b_0 c_0$ and then, since b_0, c_0 are each divisible by p, we see that p^2 must divide a_0. This means that we cannot hope to find a factorization of f in $\mathbb{Z}[x]$ unless p^2 divides a_0. In other words,

> *if a polynomial, with integer coefficients, has all its*   (1.43)
> *coefficients, except the leading one, divisible by a prime p,*
> *and if p^2 does not divide its constant coefficient,*
> *then it is irreducible in $\mathbb{Z}[x]$.*

This very powerful observation is called *Eisenstein's criterion*, after F M G Eisenstein (1823–1852).

Eisenstein's criterion can be put into another equivalent form if we first note an attractive symmetry about polynomial multiplication. For example $(2 + 3x)(7 - 5x + x^2) = 3x^3 - 13x^2 + 11x + 14$ and $(2x + 3)(7x^2 - 5x + 1) = 3 - 13x + 11x^2 + 14x^3$. This is not an accident! If we have two arbitrary polynomials $f(x) = a_k x^k + a_{k-1} x^{k-1} + \ldots + a_0$ and $g(x) = b_n x^n + b_{n-1} x^{n-1} + \ldots + b_0$, the coefficient of x^r in their product is

$$\sum_{i+j=r} a_i b_j.$$

What is the coefficient of x^r in the product of $f_1(x) = a_0 x^k + a_1 x^{k-1} + \ldots + a_k$ and $g_1(x) = b_0 x^n + b_1 x^{n-1} + \ldots + b_n$? It is easier to think about this if we write $f_1(x)$ as $A_k x^k + A_{k-1} x^{k-1} + \ldots + A_0$ and $g_1(x)$ as $B_n x^n + B_{n-1} x^{n-1} + \ldots + B_0$, where $A_i = a_{k-i}$ for each $i = 1, \ldots, k$ and $B_j = b_{n-j}$ for each $j = 1, \ldots, n$. Then the coefficient of x^r in $f_1 g_1$ is

$$\sum_{i+j=r} A_i B_j = \sum_{i+j=r} a_{k-i} b_{n-j} = \sum_{\substack{k-i+n-j \\ =k+n-r}} a_{k-i} b_{n-j}$$

and writing I for $k - i$, J for $n - j$, we get $\sum_{\substack{I+J \\ =k+n-r}} a_I b_J$ which is the

coefficient of x^{k+n-r} in fg. This is precisely what we wanted, for it
says that the product $f_1(x)g_1(x)$ is formed by taking the original product
$f(x)g(x)$ and reversing the order of the coefficients. So the fact, that
$x^3 - 6x + 7 = (x - 5)(x - 6)^2$ in $\mathbb{Z}_{17}[x]$, immediately implies that
$7x^3 - 6x^2 + 1$ is composite there and is $(1 - 5x)(1 - 6x)^2$. Likewise, the
fact that $x^6 - 4x + 2$ is irreducible in $\mathbb{Q}[x]$ implies that $2x^6 - 4x^5 + 1$ is
also irreducible in $\mathbb{Q}[x]$. In general,

> *any polynomial is composite or irreducible according as*   (1.44)
> *the polynomial with coefficients reversed is composite or irreducible.*

As a consequence in $\mathbb{Q}[x]$,

> *if a polynomial, with integer coefficients, has*   (1.45)
> *all its coefficients, except its constant term, divisible*
> *by a prime p, and if p^2 does not divide its leading coefficient,*
> *then it is irreducible in $\mathbb{Z}[x]$ and $\mathbb{Q}[x]$.*

This alternative form of Eisenstein's criterion follows from the original
version simply by applying the principle (1.44).

Activity 1.15 Which of the polynomials in Activities 1.1, 1.2, 1.3 and
1.14 (or their primitive parts) satisfy the conditions to apply Eisenstein's
criterion?

〉1.9 Factorization in $\mathbb{Q}[x]$

In $\mathbb{Q}[x]$ there are an infinite number of coefficients and so an infinite
number of possible polynomials of each degree. This makes it less certain
that we can find factors, even of quadratic polynomials, in a finite number
of steps. Notice though that if either the polynomial $\frac{a_n}{b_n}x^n + \ldots + \frac{a_0}{b_0}$
in (1.41), or its primitive part $c_n x^n + \ldots + c_0$, is divisible by a third
polynomial then so is the other. So when searching for factors of a
rational polynomial, like that in (1.41), it suffices to know how to find
factors of polynomials like $c_n x^n + \ldots + c_0$ which have integer coefficients.
We also saw, in section 1.7, that if such a polynomial has factors in $\mathbb{Q}[x]$

it must have factors in $\mathbb{Z}[x]$. So we can concentrate our efforts on finding factors in $\mathbb{Z}[x]$ of polynomials in $\mathbb{Z}[x]$.

If we have succeeded in factorizing a polynomial f with integer coefficients into a product of two other polynomials g and h with integer coefficients then we shall have an equation that looks like

$$f(x) = a_k x^k + a_{k-1} x^{k-1} + \ldots + a_0$$
$$= (b_n x^n + b_{n-1} x^{n-1} + \ldots + b_0) * (c_m x^m + c_{m-1} x^{m-1} + \ldots + c_0)$$

where $a_k, \ldots, a_0, b_n \ldots, b_0, c_m, \ldots, c_0$ are all integers. In particular the coefficients of the terms of highest degree must be the same on each side and also the constant terms must be the same. So

$$a_k = b_n c_m \text{ and } a_0 = b_0 c_0. \tag{1.46}$$

This means that the leading coefficient of each factor must divide that of f and the constant term of each factor must divide the constant term of f. Indeed, if u is any integer, we must have

$$f(u) = g(u) * h(u) \tag{1.47}$$

meaning that $g(u)$ must divide $f(u)$. To see how helpful these observations are, suppose we want to factorize the polynomial $f_1(x) = \frac{1}{5}x^3 - \frac{1}{10}x^2 + \frac{7}{10}x - \frac{3}{2}$ in $\mathbb{Q}[x]$. We first write the polynomial as $\frac{1}{10}(2x^3 - x^2 + 7x - 15)$ and then concentrate on finding factors in $\mathbb{Z}[x]$ of $f(x) = 2x^3 - x^2 + 7x - 15$. Since f here has degree 3, the inequality in (1.34) tells us that it will either be irreducible or have—at least one— linear factor. If $g(x) = ax + b$ divides f, where a and b are integers, then so does $-ax - b$ and one of a, $-a$ will be positive. We may suppose the names are chosen so that $a > 0$, and then a must be a positive divisor of the leading coefficient, 2, of f and b must divide its constant term -15. So a must be 1 or 2 and b can only be -1, 1, -3, 3, -5, 5, -15, 15. This gives sixteen possible linear factors to be examined, namely $x - 1$, $x + 1$, $x - 3$, ..., $x + 15$, $2x - 1$, $2x + 1$, ..., $2x + 15$. We can reduce the number of these further using (1.47) for, replacing x by 1, we observe that $g(1) = a + b$ must divide $f(1) = -7$. This leaves $2x - 1$, $2x - 3$ and $2x + 5$ as the only possible linear factors and attempting to divide by each in turn shows that $f(x) = (2x - 3) * (x^2 + x + 5)$. This sole linear factor of f cannot divide further into $x^2 + x + 5$ since -3 does not divide 5. So $x^2 + x + 5$ is irreducible and f_1 is therefore $\frac{1}{10}(2x - 3) * (x^2 + x + 5)$.

Activity 1.16 Where possible write each of the following polynomials in $\mathbb{Q}[x]$ as a product of irreducible factors.
(i) $x^3 - 1$; (ii) $2x^2 + 1$; (iii) $\frac{2}{3}x^3 - 3x^2 - x - \frac{10}{3}$; (iv) $x^3 - 7$; (v) $2x^3 - 2x^2 - \frac{15}{2}x + 9$.

Example 1.11
Suppose we want to factorize the polynomial $f(x) = 3x^5 - 5x^4 - 16x^3 + 56x^2 - 70x + 35$ in $\mathbb{Q}[x]$. We need only look for factors with integer coefficients and, if the polynomial is composite, it will have either linear or quadratic factors. As before, the leading coefficient of any factor g must divide 3 and its constant coefficient must divide 35. Also, by considering $-g$ instead of g if necessary, we may suppose that the leading coefficient of the factor we are looking for is positive. So the linear factors that we need to consider are $x - 1$, $x + 1$, $3x - 1$, $3x + 1$, ..., $x + 35$, $3x - 35$, $3x + 35$. But none of these is possible since $g(1)$ must divide $f(1) = 3$. We are left with the question 'does f have any quadratic factors?' Writing a supposed quadratic factor $g(x)$ as $ax^2 + bx + c$ for integers a, b, c, we have $a = 1$ or 3 and $c = -1, 1, -5, 5, -7, 7, -35$ or 35. Again $g(1) = a + b + c$ must divide $f(1) = 3$, whence

$$a + b + c = -1, 1, -3 \text{ or } 3. \tag{1.48}$$

For each of the possibilities for a and c, equation (1.48) then gives 4 possible values for b and thus $2 * 4 * 8 = 64$ possibilities for $ax^2 + bx + c$. The number of these can be further reduced by using additional relations such as the fact that $g(2)$ must divide $f(2) = 7$, which means that $4a + 2b + c = -1, 1, -7$ or 7. Eventually we divide the remaining candidates into f to see which, if any, is actually a factor. In this way we find that $f(x) = (3x^2 - 5x + 5) * (x^3 - 7x + 7)$.

In a similar way we can use the relations (1.46) and (1.47) to search for factors of any given degree of a polynomial f with integer coefficients. If we are searching for a factor g of degree n then (1.46) gives a finite number of possibilities for the leading coefficient and constant term of g. The other $n - 1$ coefficients of g are also restricted as we can see by choosing any $n - 1$ different integers $u_1, u_2, \ldots, u_{n-1}$. Then $g(u_1)$ is one of the divisors of $f(u_1)$, $g(u_2)$ is a divisor of $f(u_2)$, ..., and $g(u_{n-1})$ is a divisor of $f(u_{n-1})$. If any u_i makes $f(u_i) = 0$ then $x - u_i$ is a factor of f which we can divide out. Otherwise each of $f(u_1)$, $f(u_2)$, ..., $f(u_{n-1})$ is non-zero and has only a finite number of divisors. So there are only a finite number of possibilities for $g(u_1)$, $g(u_2)$, ..., $g(u_{n-1})$. Each set of values for $g(u_1)$, $g(u_2)$, ..., $g(u_{n-1})$ then gives one set of values for the remaining coefficients of g (just as, in (1.48), each value for $a + b + c$ gave one possible value for b). This gives a finite number of possibilities for g as a whole, and we can divide each in turn into f to see which, if any, is a factor.

The method described above is essentially due to Friedrich von Schubert in 1793 and can certainly determine the factors of any polynomial in $\mathbb{Q}[x]$ in a finite number of steps. It is not the best method of finding factors of large degree, but is easy to apply in the case of linear or quadratic factors.

〉1.10 Factorizing with the aid of the computer

The previous sections have shown that the factorization of a polynomial in either $\mathbb{Q}[x]$ or $\mathbb{Z}_p[x]$ is a finite procedure, though it can be very tedious by hand. The computer will automatically search for linear and quadratic factors of polynomials in either $\mathbb{Q}[x]$ or any domain $\mathbb{Z}_p[x]$ if you just type the name, f, g, ..., of the polynomial and press ENTER.

Example 1.12
Make sure that you are working in $\mathbb{Q}[x]$ and enter the polynomial $x^5 + 3x^4 + 4x^3 + 5x^2 + 3x + 2$ into the computer. Assuming that there are already polynomials labelled f and g, this one will be called h. Press the key 'h' (and ENTER) and the h-register should change to $(x+2)(x^2+x+1)(x^2+1)$. Type 'h' and ENTER again and the h-register will revert to $x^5 + 3x^4 + 4x^3 + 5x^2 + 3x + 2$.

It is useful to be aware of what the computer does when it is asked to factorize a polynomial. In $\mathbb{Q}[x]$ it first finds the primitive polynomial underlying the given rational polynomial and then uses the technique illustrated in Example 1.11. In a domain of the form $\mathbb{Z}_p[x]$ it searches through the p linear, and then the p^2 quadratic, possibilities which each have leading coefficient 1 (as described in section 1.8). There might be unacceptable delays on a small computer if p were larger than a few hundred so, in order to avoid that prospect, the computer restricts its quadratic searches to polynomials $x^2 + bx + c$ with $-50 \leqslant b \leqslant 50$ and $-50 \leqslant c \leqslant 50$. This means that all quadratic possibilities are examined if $p \leqslant 101$, but only a selection if $p > 101$.

Activity 1.17 Use the computer to find as many irreducible factors as possible for the following polynomials: $2x^4 + 3x^3 - 7x^2 + 8x + 7$ in $\mathbb{Z}_{17}[x]$; $x^{16} - 1$ in $\mathbb{Z}_{17}[x]$; $x^{16} - 1$ in $\mathbb{Q}[x]$; $x^8 - 1$ in $\mathbb{Q}[x]$; $x^8 + 1$ in $\mathbb{Q}[x]$; $2x^4 - 27x^3 + 27x^2 - 14x - 23$ in $\mathbb{Z}_{79}[x]$; $x^5 + 48x^4 - 133x^3 + 28x^2 + 21x + 50$ in $\mathbb{Z}_{307}[x]$ and then in $\mathbb{Z}_{101}[x]$.

If the computer has not found any (linear or quadratic) factors it will tell you, and will also tell you if that means your polynomial is irreducible. Sometimes, as mentioned just before Activity 1.7, the computer can run into overflow problems when testing possible divisors. This only happens in $\mathbb{Q}[x]$ and may mean that the computer cannot be sure that it has accounted for all the possibilities. It will tell you if that has happened and will never say that a polynomial is irreducible unless it is certain. If the computer displays some factors, you may safely conclude that any of degrees 1, 2 or 3 will be irreducible; unless, in $\mathbb{Q}[x]$, the computer has said that it has encountered numbers out of its normal range, in which case those of degrees 2 and 3 should be checked separately. Factors of degrees 4 or 5 will be irreducible if you are working in a domain $\mathbb{Z}_p[x]$ with $p \leqslant 101$, and are likely to be so in $\mathbb{Q}[x]$ though again, to be sure, you should ask the computer to check them separately. No conclusions should be made about any displayed factors of degrees 6 or more, nor about those of degrees 4 or 5 when working in a domain $\mathbb{Z}_p[x]$ with $p > 101$.

Notice that there are some cases in which a polynomial in $\mathbb{Q}[x]$ can immediately be classified as being irreducible even though it may have large degree. This is because, as we know, there are some infinite classes of polynomials that are known theoretically to be irreducible in $\mathbb{Q}[x]$. One such class consists of the polynomials whose primitive parts satisfy Eisenstein's Criterion, as discussed in section 1.8. Another consists of the *cyclotomic polynomials* which are defined in Appendix 3; and yet another is the class of polynomials, described in Appendix 4, that satisfy conditions based on Rouché's theorem. The computer knows about these classes and will immediately recognise any polynomials that belong to them.

Activity 1.18 Use the computer to check your answers to Activities 1.12, 1.13, 1.14, 1.15 and 1.16.

⟩ SUMMARY OF CHAPTER 1

In this chapter we first met polynomials with coefficients in $\mathbb{Q}$ and in $\mathbb{Z}_p$ and learned that $\mathbb{Q}[x]$ and $\mathbb{Z}_p[x]$ form integral domains. We also saw how the computer can help us to do arithmetic with polynomials and how the results of the same operations can differ when the underlying domains are changed. Much of the chapter was devoted to the problem of finding factors of polynomials. In section 1.4 we saw how in $\mathbb{Q}[x]$ and $\mathbb{Z}_p[x]$ a greatest common divisor of two polynomials can be found using Euclid's algorithm, which works in these polynomial domains just as it does among the ordinary integers. Section 1.7 introduced $\mathbb{Z}[x]$, the integral domain of all polynomials with integer coefficients. In $\mathbb{Z}[x]$ any two polynomials again have greatest common divisors, and one can always be found by using Euclid's algorithm to find a greatest common divisor in $\mathbb{Q}[x]$ and then multiplying by a suitable constant. However a GCD in $\mathbb{Z}[x]$ cannot always be expressed as a linear combination of the original two polynomials, with coefficients also in $\mathbb{Z}[x]$ (see Exercise 11 below). Euclidean domains all have that linear combination property, as is shown in general in section 4 of Appendix 1. So $\mathbb{Z}[x]$ is not a Euclidean domain.

Finding all the factors of a single polynomial is more difficult and irreducible polynomials play a central role. They are defined in the same way in each of $\mathbb{Q}[x]$, $\mathbb{Z}_p[x]$ and $\mathbb{Z}[x]$ and in all these integral domains it turns out that irreducible polynomials possess the 'prime' property (1.39) (see Exercise 9 below in the case of $\mathbb{Z}[x]$). This property then implies that every polynomial that is neither zero nor a unit is a product of irreducible polynomials in a unique way apart from order and multiplication of factors by units. The proof of this for $\mathbb{Q}[x]$ and $\mathbb{Z}_p[x]$ is given at the end of section 1.6 and the result for $\mathbb{Z}[x]$ can either be proved in a similar way or can be derived from the result for $\mathbb{Q}[x]$ (see Exercise 10). The problem of factorization in $\mathbb{Q}[x]$ can be reduced to that in $\mathbb{Z}[x]$, as explained in section 1.9; and the prime or composite nature of a polynomial in $\mathbb{Z}_p[x]$

can have immediate implications about the prime or composite nature of its counterpart in $\mathbb{Z}[x]$ (see section 1.2 and 1.8). Indeed section 1.8 shows how these connections lead to a powerful test (Eisenstein's criterion) for irreducibility in $\mathbb{Z}[x]$. Other classes of irreducible polynomials in $\mathbb{Q}[x]$ are discussed in Appendices 3 and 4, and in Exercise 14 below.

〉 EXERCISES FOR CHAPTER 1

1. Find all the irreducible cubic and quartic polynomials in $\mathbb{Z}_2[x]$.

2. Find GCDs of the following pairs of polynomials and hence or otherwise factor them completely in the indicated domains.
   (i) $3x^8 - 6x^7 + 3x^5 - 2x^4 - 6x^3 - 6x^2 - 2x + 3$ and $5x^8 - 3x^6 + 3x^5 + x^4 + 5x^3 + 2x^2 - 5x + 3$ in $\mathbb{Z}_{13}[x]$ and $\mathbb{Z}_{53}[x]$; (ii) $10x^9 + 3x^8 - 5x^7 - 18x^6 + 17x^5 - 15x^4 - 3x^3 + 5x^2 - 17x + 5$ and $2x^6 + 13x^5 + 2x^4 + 8x^3 - 16x^2 - 2x + 9$ in $\mathbb{Z}_{37}[x]$ and $\mathbb{Z}_{31}[x]$; (iii) $10x^9 + 3x^8 - 5x^7 - 18x^6 + 17x^5 - 15x^4 - 3x^3 + 5x^2 - 17x + 5$ and $x^7 - 5x^6 + 6x^5 + 11x^4 + 16x^3 - 15x^2 - 9x - 2$ in $\mathbb{Z}_{37}[x]$ and $\mathbb{Z}_{89}[x]$.

3. Find a GCD of $x^7 + x^6 - 7x^5 + 9x^4 - 2x^3 + x^2 - 1$ and $2x^8 - 4x^7 + 2x^6 + 2x^5 - 2x^4 + 5x^3 - 4x^2 - x + 1$ in each of $\mathbb{Z}_5[x]$, $\mathbb{Z}_7[x]$. Hence factorize them completely in $\mathbb{Z}_5[x]$, $\mathbb{Z}_7[x]$ and then in $\mathbb{Q}[x]$.

4. In each of the following cases find all the values of p for which the first polynomial divides the second in $\mathbb{Z}_p[x]$: $x^2 + x + 1$ and $x^5 + x^4 - 2x^3 + 4x^2 + 38x - 27$; $x^2 + x + 1$ and $x^5 + x + 1$; $2x + 7$ and $2x^3 + x^2 + x + 106$; $x^3 - x + 1$ and $2x^5 + 5x^3 + 5x^2 - 13x + 16$.

5. In any polynomial domain two polynomials are *relatively prime* if their only common divisors are units. So each unit (and 1 in particular) is a GCD of two relatively prime polynomials and is thus a linear combination of them if the domain is Euclidean. Generalise (1.39) by showing that in a Euclidean domain if a polynomial divides a product fg and is relatively prime to f then it must divide g. Use this to show that if a polynomial is divisible by each of two relatively prime polynomials, then it must be divisible by their product.

6. Let f, h_1, h_2 be polynomials in $F[x]$ for some field F, with h_1 and h_2 relatively prime. Suppose that G is a GCD of f and $h_1 h_2$, and that g_1, g_2 are GCDs of f, h_1 and f, h_2 respectively. Use the result of the last question to prove that $g_1 g_2$ divides G. Then by writing g_1, g_2 as linear combinations, prove that G divides $g_1 g_2$. Deduce that $g_1 g_2$ is a GCD of f and $h_1 h_2$.

7. Let f be a polynomial in $\mathbb{Z}[x]$ which is primitive. If f is irreducible in $\mathbb{Q}[x]$ prove that it must also be irreducible in $\mathbb{Z}[x]$. Deduce that, in $\mathbb{Q}[x]$, each irreducible polynomial is a constant multiple of an irreducible polynomial in $\mathbb{Z}[x]$.

8. Let f and g be polynomials in $\mathbb{Z}[x]$ with $\deg(f) \geqslant 1$ and $\deg(g) \geqslant 1$. Suppose that g is primitive and that $f = qg$ with $q \in \mathbb{Q}[x]$. Prove that q must actually be in $\mathbb{Z}[x]$.

9. Prove that a polynomial p that is irreducible in $\mathbb{Z}[x]$ must have the prime property in $\mathbb{Z}[x]$. [In the case of $\deg(p) \geqslant 1$ use the fact that p will be irreducible and thus prime in $\mathbb{Q}[x]$.]

10. Prove that non-zero and non-unit polynomials in $\mathbb{Z}[x]$ are products of irreducible polynomials in $\mathbb{Z}[x]$ in essentially only one way.

11. Prove that, in $\mathbb{Z}[x]$, 1 is a GCD of $10x$ and $x^2 + 1$. Show further that there cannot be any polynomials r, s in $\mathbb{Z}[x]$ with $1 = r(x)(x^2 + 1) + 10xs(x)$. This shows that $\mathbb{Z}[x]$ is not a Euclidean domain (see section 4 in Appendix 1).

12. Suppose that the integer n is at least 2 and that F is any field. Prove that the polynomial $f(x) = x^{n-1} + x^{n-2} + \ldots + x + 1$ can only be irreducible in $F[x]$ if n is prime. The converse, that f is irreducible if n is prime, is true if $F = \mathbb{Q}$ (see Appendix 3). Is the converse always true if F is a finite field?

13. If f is a polynomial in $\mathbb{Z}[x]$ show that a rational number $\frac{a}{b}$ can only be a root of f if every prime dividing b also divides the leading coefficient of f. Deduce that if f is monic then its rational roots must be integers.

14. Suppose that $f(x) = a_{2k+1} x^{2k+1} + a_{2k} x^{2k} + \ldots + a_1 x + a_0$ and that

the odd degree of f is at least 3. Suppose further that there is a prime p that divides every coefficient of f except the leading one; that $p^2 | a_i$ for $0 \leqslant i \leqslant k$; and that p^3 does not divide a_0. Follow the analysis of section 1.8 to establish that f must be irreducible. This result is due to E Netto in 1897.

〉 Chapter 2

〉 Using polynomials to make new number fields

〉2.1 Roots of irreducible polynomials

Consider a polynomial $f(x)$ of degree greater than 1 whose coefficients lie in either the rational field $\mathbb{Q}$ or a field $\mathbb{Z}_p$ for some prime p. For convenience we shall let F stand for either $\mathbb{Q}$ or the appropriate $\mathbb{Z}_p$, so that we can write $f(x) \in F[x]$. Now suppose that f is irreducible in $F[x]$. Then f cannot have any roots in F (for the existence of a root α in F would of course imply that f had a linear factor $x - \alpha$ in $F[x]$); but there is no reason why f could not have a root in some larger field than F.

For example, we know from Eisenstein's criterion that the polynomial $x^2 - 2$ is irreducible in $\mathbb{Q}[x]$ and so has no roots in $\mathbb{Q}$. However it has two roots, $\sqrt{2}$ and $-\sqrt{2}$, in the larger field $\mathbb{R}$ of real numbers and factorizes in $\mathbb{R}[x]$ as $(x - \sqrt{2})(x + \sqrt{2})$. We say that $\mathbb{R}$ is an *extension field* of $\mathbb{Q}$ or, equivalently, that $\mathbb{Q}$ is a *subfield* of $\mathbb{R}$. Actually we do not have to bring in all the real numbers to find roots of $x^2 - 2$. We shall see shortly that an intermediate field containing $\mathbb{Q}$, and contained in $\mathbb{R}$, is formed by the set of all numbers of the form $a + b\sqrt{2}$ where a and b are rational. This set is denoted by $\mathbb{Q}(\sqrt{2})$ and contains the whole of $\mathbb{Q}$ since every rational number a can be written as $a + 0\sqrt{2}$ — a rational plus a rational multiple of $\sqrt{2}$. It also contains $\sqrt{2}$ and $-\sqrt{2}$ since they can similarly be expressed as $0 + 1\sqrt{2}$ and $0 + (-1)\sqrt{2}$ respectively. In order to see that $\mathbb{Q}(\sqrt{2})$ forms a field with respect to ordinary addition and multiplication we note that its elements are particular sorts of real numbers, and so the commutative and associative properties of addition and multiplication automatically hold for them as they do for all real numbers. Likewise multiplication is distributive with respect to addition in $\mathbb{Q}(\sqrt{2})$ and the additive and multiplicative identities 0 and 1 are in $\mathbb{Q}(\sqrt{2})$. The only properties that need to be specially checked are that $\mathbb{Q}(\sqrt{2})$ is closed under addition and multiplication of its elements; and also that every element $a + b\sqrt{2}$ has both an additive and a multiplicative inverse of the same form. As regards addition, if $a + b\sqrt{2}$ and $c + d\sqrt{2}$ are in $\mathbb{Q}(\sqrt{2})$ then so is their sum, since

$$(a + b\sqrt{2}) + (c + d\sqrt{2}) = (a + c) + (b + d)\sqrt{2} \qquad (2.1)$$

is again a rational plus a rational multiple of $\sqrt{2}$; and

$$-(a + b\sqrt{2}) = -a - b\sqrt{2} \qquad (2.2)$$

is clearly also in $\mathbb{Q}(\sqrt{2})$. Similarly every product of elements of $\mathbb{Q}(\sqrt{2})$ is in the same set since

$$(a + b\sqrt{2}) * (c + d\sqrt{2}) = (ac + 2bd) + (ad + bc)\sqrt{2}. \qquad (2.3)$$

The multiplicative inverse of $a + b\sqrt{2}$ is less obvious, but

$$(a + b\sqrt{2}) * (a - b\sqrt{2}) = a^2 - 2b^2 \qquad (2.4)$$

and, because $\sqrt{2}$ is irrational, we could not have $2 = a^2/b^2$ so that $a^2 - 2b^2$ could only be zero if both a and b were zero (or, in other words, if $a + b\sqrt{2}$ were zero). Thus if $a + b\sqrt{2}$ is not zero then

$$\frac{1}{a + b\sqrt{2}} = \frac{a - b\sqrt{2}}{a^2 - 2b^2} = \frac{a}{a^2 - 2b^2} - \left(\frac{b}{a^2 - 2b^2}\right)\sqrt{2} \qquad (2.5)$$

which is again a rational plus a rational multiple of $\sqrt{2}$.

Example 2.1
A particular case of (2.4) is the equation $(5 + 3\sqrt{2}) * (5 - 3\sqrt{2}) = 7$, which implies

$$\frac{1}{5 + 3\sqrt{2}} = \frac{5 - 3\sqrt{2}}{(5 + 3\sqrt{2})(5 - 3\sqrt{2})} = \frac{5}{7} - \frac{3}{7}\sqrt{2}$$

in accordance with (2.5).

The number $a^2 - 2b^2$, which is the product of $a + b\sqrt{2}$ and $a - b\sqrt{2}$ in (2.4), also arises in the product of $a + bx$ and $a - bx$ if we look at this polynomial product† modulo $x^2 - 2$:

$$(a + bx) * (a - bx) = a^2 - b^2 x^2 = -b^2(x^2 - 2) + a^2 - 2b^2 \quad (2.6)$$
$$\equiv a^2 - 2b^2 \pmod{x^2 - 2}.$$

So, modulo $x^2 - 2$, it makes sense to think of $a + bx$ as having the multiplicative inverse $\frac{1}{a^2-2b^2}(a - bx)$ since, whenever $a + bx \neq 0$, (2.6) implies

$$(a + bx) * \left\{ \frac{a - bx}{a^2 - 2b^2} \right\} \equiv 1 \pmod{x^2 - 2}. \tag{2.7}$$

Working modulo $x^2 - 2$ we can also reproduce the coefficients in (2.3):

$$(a + bx)(c+dx) = ac + (bc + ad)x + bdx^2$$
$$= (ac + 2bd) + (bc + ad)x + bd(x^2 - 2)$$
$$\equiv (ac + 2bd) + (bc + ad)x \pmod{x^2 - 2}. \tag{2.8}$$

Constructing analogues of (2.1) and (2.2) in the same vein is even more straightforward, since

$$(a+bx)+(c+dx) = (a+c)+(b+d)x \equiv (a+c)+(b+d)x \pmod{x^2-2} \tag{2.9}$$

and

$$-(a + bx) = -a - bx \equiv -a - bx \pmod{x^2 - 2}. \tag{2.10}$$

In other words, by doing arithmetic with linear and constant polynomials and working modulo $x^2 - 2$, we can mirror essential properties of expressions involving $\sqrt{2}$, without ever mentioning $\sqrt{2}$ itself! All the other arithmetic properties of $\mathbb{Q}(\sqrt{2})$ are also present in our polynomial congruences. For example, addition and multiplication of polynomials f and g remains commutative: $f + g \equiv g + f \pmod{x^2 - 2}$ and $fg \equiv gf \pmod{x^2 - 2}$. Similarly polynomial addition and multiplication

---

† By analogy with congruences in $\mathbb{Z}$, we say that polynomials f, g are congruent modulo h and write $f \equiv g \pmod{h}$, whenever $f - g$ is divisible by h.

are associative and multiplication is distributive with respect to addition when considered modulo $x^2 - 2$. The numbers 0 and 1, which are additive and multiplicative identities in $\mathbb{Q}[x]$, still play those roles modulo $x^2 - 2$ because $0 * f \equiv 0 \pmod{x^2 - 2}$ and $1 * f \equiv f \pmod{x^2 - 2}$. What this means is that the set of all linear and constant polynomials, with rational coefficients, forms a field with respect to the operations of addition modulo $x^2 - 2$ and multiplication modulo $x^2 - 2$. This field is denoted by $\mathbb{Q}[x]/\langle x^2 - 2 \rangle$, and in it the arithmetic properties of the element $a + bx$ exactly reflect those of the element $a + b\sqrt{2}$ in $\mathbb{Q}(\sqrt{2})$.

Example 2.2
The criterion that defines a positive real number as $\sqrt{2}$ is that its square is 2. In $\mathbb{Q}[x]/\langle x^2 - 2 \rangle$ the element corresponding to $\sqrt{2} = 0 + 1\sqrt{2}$ is $0 + 1x = x$ and indeed we find that

$$x^2 = 2 + (x^2 - 2) \equiv 2 \pmod{x^2 - 2}.$$

Every property that can be found in $\mathbb{Q}(\sqrt{2})$ can also be found in $\mathbb{Q}[x]/\langle x^2 - 2 \rangle$ and vice versa. So, as far as arithmetic properties are concerned, $\mathbb{Q}[x]/\langle x^2 - 2 \rangle$ is an exact copy of $\mathbb{Q}(\sqrt{2})$, the only difference being that the elements and operations have different names. This situation is described by saying that the two fields are *isomorphic*. It may seem a little perverse to go to the trouble of replacing $\mathbb{Q}(\sqrt{2})$ by another field that is arithmetically identical; but the point is that we have constructed $\mathbb{Q}[x]/\langle x^2 - 2 \rangle$ without using any previously known roots of $x^2 - 2$. What is more, the element x in our new field has every right to be called a 'square root of 2' since, as Example 2.2 shows, when x is combined with itself in $\mathbb{Q}[x]/\langle x^2 - 2 \rangle$ the result is 2. In other words, in $\mathbb{Q}[x]/\langle x^2 - 2 \rangle$, when the operations of squaring and then subtracting 2 are applied to x we get zero. This means that if we create a new variable, y say, which can range over all the elements of the field $K = \mathbb{Q}[x]/\langle x^2 - 2 \rangle$ then we can think of x as being a root of the polynomial $y^2 - 2$. So just starting from $\mathbb{Q}$ and the original polynomial $x^2 - 2 \in \mathbb{Q}[x]$ we have constructed a field K that contains a root of the analogous polynomial $y^2 - 2 \in K[y]$. This new field contains every constant polynomial, which is to say every rational number. So it is an extension field of $\mathbb{Q}$ just as $\mathbb{Q}(\sqrt{2})$ is.

The computer can help to do arithmetic in $\mathbb{Q}[x]/\langle x^2 - 2 \rangle$ by taking care of any awkward modular calculations. Make sure that you are working with rational polynomials and then enter '$f = x^2 - 2$'. After typing 'f'

(and 'ENTER') the computer will of course say that f is irreducible and offers the chance of just pressing 'ENTER' by itself to investigate the extension field $\mathbb{Q}[x]/\langle x^2 - 2 \rangle$. When this is done notice that it replaces f by 0 since $x^2 - 2 \equiv 0 \pmod{x^2 - 2}$. It also reduces any other polynomials modulo $x^2 - 2$. Now try entering '$f = x - 1$' and '$g = x + 3$' and ask for $f * g$. Ordinarily we should expect the product of $x - 1$ and $x + 3$ to be $x^2 + 2x - 3$, but the computer says that the product of f and g is $2x - 1$ because $x^2 + 2x - 3 \equiv 2x - 1 \pmod{x^2 - 2}$. The result of '$f/g$' is even less like the result would be in $\mathbb{Q}[x]$. There is no remainder since division by non-zero polynomials is always exact here (as is to be expected in a field), and the equation $\frac{x-1}{x+3} = \frac{4}{7}x - \frac{5}{7}$ in $\mathbb{Q}[x]/\langle x^2 - 2 \rangle$ is equivalent to the congruence

$$x - 1 \equiv (x + 3) * (\tfrac{4}{7}x - \tfrac{5}{7}) \pmod{x^2 - 2}$$

in $\mathbb{Q}[x]$.

Activity 2.1 For each of the following pairs of polynomials f and g, find $f * g$ and f/g in $\mathbb{Q}[x]/\langle x^2 - 2 \rangle$.
  (i) $f = x + 3$, $g = 5x - \frac{3}{4}$; (ii) $f = x + 3$, $g = 3x^3 + \frac{2}{3}x^2 - 7x + 2$;
  (iii) $f = 2x - 5$, $g = x - 1$; (iv) $f = 2x - 5$, $g = x + 1$.

The same method of using polynomial congruences to construct extension fields can be used in less familiar situations. In $\mathbb{Z}_3[x]$, for instance, the polynomial $x^2 - 2$ is again irreducible and there is no obvious extension field K of $\mathbb{Z}_3$ that might contain roots for the analogous polynomial $y^2 - 2 \in K[y]$. However the congruences (2.7)–(2.10) still hold good when the coefficients are reduced modulo 3, or in other words when the polynomials are interpreted as elements of $\mathbb{Z}_3[x]$. The commutative and associative properties and the distributivity of multiplication with respect to addition are also true when we interpret addition as 'addition of polynomials in $\mathbb{Z}_3[x]$ modulo $x^2 - 2$' and multiplication as 'multiplication of polynomials in $\mathbb{Z}_3[x]$ modulo $x^2 - 2$'. So the set of constant and linear polynomials in $\mathbb{Z}_3[x]$ with these operations modulo 3 *and* modulo $x^2 - 2$ forms a field called $\mathbb{Z}_3[x]/\langle x^2 - 2 \rangle$. Again the computer can help with the calculations here.

Activity 2.2 Make sure that you are using polynomial arithmetic in $\mathbb{Z}_3[x]$ and enter the polynomial $x^2 - 2$ (which will appear as $x^2 + 1$). Type the name that the computer has given to the polynomial and press 'ENTER' to test it for factors. The computer will say that it is irreducible and invite you to press 'ENTER' again to go into the extension field $\mathbb{Z}_3[x]/\langle x^2 - 2\rangle$. Do this and then enter '$f = x$' and '$g = 2$'. First calculate $f * f$ and then $h - g$. What do you get? You have just shown that, in this extension field of $\mathbb{Z}_3$, x is indeed a root of the polynomial that says 'square and then subtract 2'.

Activity 2.3 For each of the following pairs of polynomials f and g, find $f + g$, $f * g$ and f/g in $\mathbb{Z}_3[x]/\langle x^2 - 2\rangle$.
(i) $f = x - 1$, $g = x + 3$; (ii) $f = x - 1$, $g = 2x - 5$; (iii) $f = x + 1$, $g = x$.

We can actually list all the elements of $\mathbb{Z}_3[x]/\langle x^2 - 2\rangle$ because they are all of the form $a + bx$ where a and b are elements of $\mathbb{Z}_3$. Since there are only three elements in $\mathbb{Z}_3$, namely 0, 1 and -1, there are only nine elements in $\mathbb{Z}_3[x]/\langle x^2 - 2\rangle$ and they are

$$0,\ 1,\ -1,\ x,\ 1+x,\ -1+x,\ -x,\ 1-x,\ -1-x.$$

Example 2.3
The number $\sqrt{5}$ is irrational and, just as for $\mathbb{Q}(\sqrt{2})$, the set of elements $a + b\sqrt{5}$ for a and b rational, form a field denoted by $\mathbb{Q}(\sqrt{5})$. This is because the equations (2.1)–(2.5) still hold when 2 and $\sqrt{2}$ are replaced throughout by 5 and $\sqrt{5}$ respectively. Similarly the congruences (2.7)–(2.10) are true if 2 is replaced in each of them by 5, and then as before the set of linear and constant polynomials in $\mathbb{Q}[x]$ forms a field with the operations of addition modulo $x^2 - 5$ and multiplication modulo $x^2 - 5$. This field $\mathbb{Q}[x]/\langle x^2 - 5\rangle$ is isomorphic to $\mathbb{Q}(\sqrt{5})$. Similarly, if $\sqrt{d}$ is irrational for an integer d, we could replace 2 and $\sqrt{2}$ in (2.1)–(2.10) by d and $\sqrt{d}$ and see that the set of elements $a + b\sqrt{d}$, for a and b rational, form the field $\mathbb{Q}(\sqrt{d})$ which is isomorphic to the field $\mathbb{Q}[x]/\langle x^2 - d\rangle$ of linear and constant polynomials in $\mathbb{Q}[x]$ with the operations of 'addition modulo $x^2 - d$' and 'multiplication modulo $x^2 - d$'.

Example 2.4
Suppose, for some prime p, that d is an element of $\mathbb{Z}_p$ which is not a square in that field. This is equivalent to saying that $x^2 - d$ is irreducible in $\mathbb{Z}_p[x]$ and, as in the case of $x^2 - 2 \in \mathbb{Z}_3[x]$, the set of all linear and constant polynomials in $\mathbb{Z}_p[x]$, with the operations of addition of polynomials in $\mathbb{Z}_p[x]$ modulo $x^2 - d$ and multiplication of polynomials in $\mathbb{Z}_p[x]$ modulo $x^2 - d$, forms a field $\mathbb{Z}_p[x]/\langle x^2 - d\rangle$.

Activity 2.4 For each of the choices of d and p below, use the computer to examine the polynomial $x^2 - d$ in $\mathbb{Z}_p[x]$ and thus decide whether d is a square in $\mathbb{Z}_p$. (In Chapter 4 we shall see a way of testing whether d is a square element in $\mathbb{Z}_p$ without involving $x^2 - d$.)
  (i) $d = 2$, $p = 17$; (ii) $d = 2$, $p = 29$; (iii) $d = -7$, $p = 419$;
  (iv) $d = 93$, $p = 211$; (v) $d = -336$, $p = 1093$;
  (vi) $d = 29$, $p = 97$.
In each case where d is a square look at the linear factors of $x^2 - d$ in $\mathbb{Z}_p[x]$ and so name a square root of d in $\mathbb{Z}_p$.

The whole process of forming an extension field as above goes through practically unchanged for any irreducible polynomial either in $\mathbb{Q}[x]$ or in one of the domains $\mathbb{Z}_q[x]$. In order to see this, suppose that $p(x)$ is irreducible in $F[x]$ where F is either $\mathbb{Q}$ or $\mathbb{Z}_q$ for some prime q. Following our previous discussions we shall look at the set of polynomials in $F[x]$ which are either constants or have degrees less than that of $p(x)$. We can certainly call this set $F[x]/\langle p(x) \rangle$ without yet implying anything by that choice of name. Addition and subtraction modulo p are defined in this set just as ordinary polynomial addition and subtraction. The set is closed with respect to these operations since, if f and g are in the set, then so are $f + g$ and $-g$ as they too have degrees less than $\deg(p(x))$. It is also closed with respect to the operation of multiplication modulo $p(x)$, because we know that we can divide fg by $p(x)$ leaving a unique remainder $h(x)$ of smaller degree than $p(x)$ and then have $fg \equiv h$ (mod p) where h is in $F[x]/\langle p(x) \rangle$. The polynomial h is 'the product of f and g modulo p'. Again polynomial addition and multiplication are commutative and associative and multiplication is distributive with respect to addition when considered modulo $p(x)$. The numbers 0 and 1 are also the identities for addition and multiplication in $F[x]/\langle p(x) \rangle$ since $0 * f = 0 \equiv 0$ (mod $p(x)$) and $1 * f = f \equiv f$ (mod $p(x)$). It is a little more difficult to see that any non-zero polynomial f in $F[x]/\langle p(x) \rangle$ has a multiplicative inverse; but such an f, being of smaller degree than the irreducible polynomial p, must be relatively prime to it. This means that there have to be polynomials u and v in $F[x]$ such that

$$u(x)f(x) + v(x)p(x) = 1 \qquad (2.11)$$

just as in (1.29). In (2.11) the polynomial multiplier of $f(x)$ can be taken to be of smaller degree than $p(x)$ by expressing it, if necessary, as $u(x) = q(x)p(x) + u_1(x)$ where $\deg(u_1) < \deg(p)$ so that (2.11) becomes $u_1(x)f(x) + [v(x) + q(x)f(x)]p(x) = 1$. Suppose that this has already been done and that $u(x)$ in (2.11) has degree less than $p(x)$. Then

$$u(x)f(x) \equiv 1 \pmod{p(x)}, \tag{2.12}$$

which is the same as

$$u(x)f(x) = 1 \text{ in } F[x]/\langle p(x) \rangle. \tag{2.13}$$

Thus $u(x)$ is the inverse of $f(x)$ and it follows that $F[x]/\langle p(x) \rangle$ is a field with respect to the operations of 'polynomial addition modulo $p(x)$' and 'polynomial multiplication modulo $p(x)$'. Note that this field contains all the constant polynomials, which are the elements of F. So it is always an extension field of F.

If the degree of p is n then each element of $F[x]/\langle p(x) \rangle$ must be of the form $a_{n-1}x^{n-1} + a_{n-2}x^{n-2} + \ldots + a_0$. In other words every element is a linear combination of the n powers $x^0, x^1, \ldots, x^{n-1}$, with coefficients in F. We say that the field $F[x]/\langle p(x) \rangle$ has *degree n* over F. If $F = \mathbb{Z}_q$ there are only q choices for each coefficient and so q^n possible linear combinations. So, if there is an irreducible polynomial $p(x)$ of degree n in $\mathbb{Z}_q[x]$, the extension field $\mathbb{Z}_q[x]/\langle p(x) \rangle$ will contain q^n elements.

Activity 2.5 In $\mathbb{Q}[x]$ verify that the polynomial $x^3 + x + 1$ is irreducible and, for each of the following pairs of polynomials f and g, find $f + g$, $f * g$ and f/g in $\mathbb{Q}[x]/\langle x^3 + x + 1 \rangle$.
(i) $f = x^2 - 1$, $g = x + 3$; (ii) $f = x^2 + 1$, $g = x$;
(iii) $f = \frac{2}{3}x^2 + x - 3$, $g = 4x^2 + 3x - 1$.

Activity 2.6 In $\mathbb{Z}_{17}[x]$ verify that the polynomial $x^4 + x^3 + x^2 + x + 1$ is irreducible and, for each of the following pairs of polynomials f and g, find $f + g$, $f * g$ and f/g in $\mathbb{Z}_{17}[x]/\langle x^4 + x^3 + x^2 + x + 1 \rangle$.
(i) $f = -5x^2 + x - 3$, $g = 4x^2 + 3x - 1$; (ii) $f = x^2 + 1$, $g = x$;
(iii) $f = 2x^3 + 6x - 1$, $g = 8x^2 - 3x$.

Notice what would happen if we tried to use a similar construction modulo a composite polynomial. Suppose for instance that in $\mathbb{Q}[x]$ we considered all the polynomials of degree less than 3 and added and multiplied them modulo $x^3 + 2x^2 + x + 2 = (x^2 + 1)(x + 2)$. That is perfectly legitimate and the resulting algebraic structure is a ring; but it is not a field. In this example that is because neither $x^2 + 1$ nor $x + 2$ is zero in the ring but their product is zero as

$$(x^2 + 1) * (x + 2) \equiv 0 \pmod{x^3 + 2x^2 + x + 2}.$$

In the ring $\mathbb{Q}[x]/\langle x^3 + 2x^2 + x + 2 \rangle$ this congruence amounts to $(x^2 + 1) * (x + 2) = 0$; so, if $x^2 + 1$ had an inverse $u(x)$ say, then $0 = u(x) * (x^2 + 1) * (x + 2) = 1 * (x + 2) = x + 2$, whereas $x + 2$ is not zero. Similarly, if we started from the field F (either $\mathbb{Q}$ or $\mathbb{Z}_q$) and used congruences modulo any composite polynomial $f(x) = a(x)b(x)$, neither of the two factors $a(x)$, $b(x)$ would have multiplicative inverses in $F[x]/\langle f(x) \rangle$. From now on we shall only use the notation $F[x]/\langle p(x) \rangle$ with an irreducible polynomial $p(x)$.

No matter what field F and irreducible polynomial $p(x)$ we start from, the extension field $F[x]/\langle p(x) \rangle$ always contains a root of the polynomial p. This is because the obvious congruence

$$p(x) \equiv 0 \; (\mathrm{mod}\; p(x))$$

in $F[x]$ is equivalent to the equation $p(x) = 0$ in $K = F[x]/\langle p(x) \rangle$; and that means that x is a root of the polynomial $p(y)$ in $K[y]$. Activity 2.2 illustrated this in the case of $F = \mathbb{Z}_3$ and $p(x) = x^2 - 2$, and for any irreducible p you can similarly use the computer to evaluate $p(x)$ in $F[x]/\langle p(x) \rangle$. You should always get $p(x) = 0$.

Activity 2.7 In Activity 2.4 you should have found that d is not a square in $\mathbb{Z}_p$ for the following choices of d and p : $d = 2$, $p = 29$; $d = -7$, $p = 419$; $d = 29$, $p = 97$. In each of these cases use the computer to enter the extension field $\mathbb{Z}_p[x]/\langle x^2 - d \rangle$, and then multiply x by itself to verify that x is a 'square root' of d in this field.

Once an extension field containing a root of the irreducible polynomial p has been constructed, it often simplifies discussions to refer to the root as α and the smallest field containing it as $F(\alpha)$, to be read as 'F extended by α'. As in the remarks after (2.13), if the degree of p is n then every element of $F(\alpha)$ is of the form $a_{n-1}\alpha^{n-1} + a_{n-2}\alpha^{n-2} + \ldots + a_0$ with coefficients from F; and if F is finite with q elements then $F(\alpha)$ has q^n elements.

$\rangle$2.2 Splitting fields

Given any polynomial f, composite or irreducible, with coefficients in a field F, we can now see that there will be a field K that contains roots of f. If f has factors of degree 1 in $F[x]$ then we can take $K = F$.

Otherwise f will be a product of irreducible factors $p_1, \ldots, p_t$ (where $t = 1$ if f is irreducible) and we can take $K = F[x]/\langle p_1(x)\rangle$ which will contain a root of p_1 and so of f.

It is quite possible for a polynomial to have more than one root in a field. For instance, the polynomial $x^2 - 2$ has the two roots $\sqrt{2}$ and $-\sqrt{2}$ in $\mathbb{Q}(\sqrt{2})$ (or equivalently, the polynomial $y^2 - 2$ has the two roots x and $-x$ in $\mathbb{Q}[x]/\langle x^2 - 2\rangle$). However it can never have more roots than its degree. This is because of the general result that a polynomial of degree $n \geqslant 1$ can have at most n roots in any field containing its coefficients. The easiest way to see this is to use induction on the degree of the polynomial. It is certainly true for linear polynomials as any linear polynomial $ax + b$ has degree 1 and also has one root $-\frac{b}{a}$. Suppose it is true for any polynomial of degree less than r, for some $r > 1$, and then consider a polynomial $g(x)$ of degree r. If g has no roots in the specified field then the result is true. If it has at least one root, $x = \alpha$ say, then, as in chapter 1, section 1.5, we can divide $g(x)$ by $x - \alpha$ obtaining

$$g(x) = (x - \alpha)h(x)$$

where $h(x)$ is a polynomial of degree $r - 1$. Apart from α all roots of g have to be roots of h. But since h has degree $r - 1$ it can have at most $r - 1$ roots. So g has at most $1 + (r - 1) = r$.

It is of course also possible for a polynomial of degree $n > 1$ to have fewer than n roots in a field containing its coefficients (and possibly none at all). Say $f(x) \in F[x]$ has degree n and has roots $\alpha_1, \alpha_2, \ldots, \alpha_k$ in F where $k < n$. Then in $F[x]$

$$f(x) = (x - \alpha_1) \ldots (x - \alpha_k)g(x) \tag{2.14}$$

where $g(x) \in F[x]$ has no linear factors and so is of degree 2 or more. As in the first paragraph of this section, we can construct an extension field K of F containing a root of g. Since it contains F, the new field K will contain all the previous roots $\alpha_1, \alpha_2, \ldots, \alpha_k$ as well as at least one root, say α_{k+1}, of g (which will also be a root of f). If g does not have as many roots in K as its degree allows, then in $K[x]$ the polynomial $g(x)$ will factorize as

$$g(x) = (x - \alpha_{k+1}) \ldots h(x) \tag{2.15}$$

where $h(x) \in K[x]$ has no linear factors. This implies

$$f(x) = (x - \alpha_1) \ldots (x - \alpha_k)(x - \alpha_{k+1}) \ldots h(x) \tag{2.16}$$

and as before we can construct a still larger† field that will contain at least one root of h, and so another root of f. If we continue constructing larger fields that contain more and more roots of f then after fewer than n steps we shall have constructed a field L that contains n roots $\alpha_1, \alpha_2, \ldots, \alpha_n$ of f. In $L[x]$ the polynomial f then splits completely into linear factors

$$f(x) = (x - \alpha_1) \ldots (x - \alpha_n). \tag{2.17}$$

If $f(x) \in F[x]$, the smallest extension field of F in which f splits into n linear factors is called the *splitting field* of f with respect to F.

For example the polynomial $f_1(y) = y^2 - 2$ with coefficients in $\mathbb{Q}$ has the splitting field $\mathbb{Q}(\sqrt{2})$ since that is the smallest extension field of $\mathbb{Q}$ that contains two roots of the polynomial. We know that the field $\mathbb{Q}[x]/\langle x^2 - 2 \rangle$ also contains two roots of f_1 and since it is isomorphic to $\mathbb{Q}(\sqrt{2})$ it can also be regarded as a splitting field of f_1. If f is any polynomial with coefficients in a field F, it is known that if two fields K_1 and K_2 are both extension fields of F and are both splitting fields of f then they have to be isomorphic (see e.g. Allenby(1991)).

⟩2.3 Some easy binomial theorems in $\mathbb{Z}_p$

In any ring or field the usual binomial theorem holds. So if a and b are elements of the ring or field and n is a positive integer then

$$(a + b)^n = a^n + na^{n-1}b + \frac{n(n-1)}{2}a^{n-2}b^2 + \ldots + \binom{n}{i} a^{n-i}b^i + \ldots + b^n \tag{2.18}$$

where the integer coefficient $\binom{n}{i} = \frac{n!}{i!(n-i)!}$ is the number of ways of choosing i objects from n. In the field $\mathbb{Z}_p$, for a prime p, there are some cases that are extremely simple, and the first of these is the case when $n = p$. In this case the expansion (2.18) becomes

$$(a + b)^p = a^p + pa^{p-1}b + \ldots + \binom{p}{i} a^{p-i}b^i + \ldots + b^p \tag{2.19}$$

and the coefficient of $a^{p-i}b^i$ in (2.19) is

$$\frac{p!}{i!(p-i)!} = \frac{1.2\ldots p}{(1\ldots i)(1\ldots p-i)}.$$

† A field L' is smaller (respectively larger) than a field L if $L' \subset L$ ($L \subset L'$).

For $1 \leqslant i < p$ all of the numbers in the denominator are smaller than p and so, as p is prime, have no factors in common with it. Thus the integer that is left when $i!(p - i)!$ is completely divided into $p!$ must still have a factor p. So, for $1 \leqslant i < p$,

$$\binom{p}{i} \equiv 0 \pmod{p}. \tag{2.20}$$

Thus the expansion (2.19) implies

$$(a + b)^p \equiv a^p + b^p \pmod{p} \tag{2.21}$$

which in $\mathbb{Z}_p$ is the same as

$$(a + b)^p = a^p + b^p. \tag{2.22}$$

Example 2.5
Modulo 5 we have $(3+4)^5 \equiv 2^5 \equiv 2$; $3^5 = 243 \equiv 3$; and $4^5 = 1024 \equiv 4$. Thus $(3+4)^5 \equiv 3^5 + 4^5 \pmod{5}$ and $(3+4)^5 = 3^5 + 4^5$ in $\mathbb{Z}_5$.

The other simple binomial expansions in $\mathbb{Z}_p$ occur when the exponent is any power of p. Applying (2.22) twice, for instance, results in

$$(a + b)^{p^2} = ((a + b)^p)^p = (a^p + b^p)^p = a^{p^2} + b^{p^2}. \tag{2.23}$$

Similarly, after k applications of (2.22), we see that

$$(a + b)^{p^k} = \left((a + b)^{p^{k-1}}\right)^p = \left(a^{p^{k-1}} + b^{p^{k-1}}\right)^p = a^{p^k} + b^{p^k}. \tag{2.24}$$

The equations (2.22), (2.23) and (2.24) are not only true in $\mathbb{Z}_p$. They also hold in any ring or field that contains $\mathbb{Z}_p$. This is because p (strictly the corresponding ring element that is the sum of p copies of the multiplicative identity) will again be zero there as it is in $\mathbb{Z}_p$. So if c is an element of such a ring or field then pc will be zero since $pc = c + c + \ldots + c = c(1 + 1 + \ldots + 1)$ and indeed pk copies of c will again be zero for any integer k. So if in an extension of $\mathbb{Z}_p$ we expand $(a + b)^p$ as in (2.19) then, as before, the binomial coefficients $\binom{p}{i}$ will all vanish for $1 \leqslant i < p$ and we again get (2.22).

Example 2.6
The ring $\mathbb{Z}_p[x]$ contains $\mathbb{Z}_p$ and so the equations (2.22), (2.23) and (2.24) all hold when a and b are elements of $\mathbb{Z}_p[x]$. In particular, using $a = x$ and $b = -1$ in (2.22),

$$(x - 1)^p = x^p + (-1)^p = x^p - 1 \text{ in } \mathbb{Z}_p[x]. \qquad (2.25)$$

(Note that when $p = 2$ we have $(-1)^2 = 1 = -1$ in $\mathbb{Z}_2$ so (2.25) still holds as written.)

Activity 2.8 Factorize (i) $x^2 - 1$ in $\mathbb{Z}_2[x]$, $\mathbb{Z}_3[x]$; (ii) $x^3 - 1$ in $\mathbb{Z}_2[x]$, $\mathbb{Z}_3[x]$, $\mathbb{Z}_7[x]$; (iii) $x^5 - 1$ in $\mathbb{Z}_2[x]$, $\mathbb{Z}_3[x]$, $\mathbb{Z}_5[x]$, $\mathbb{Q}[x]$.

Each of equations (2.22), (2.23) and (2.24) can be extended to apply to a sum of several terms. Thus, from (2.22) for example,

$$(a_1 + a_2 + \ldots + a_r)^p = (a_1 + a_2 + \ldots + a_{r-1})^p + a_r^p = \ldots$$
$$= a_1^p + a_2^p + \ldots + a_r^p \qquad (2.26)$$

for any $a_1, a_2, \ldots, a_r$ in $\mathbb{Z}_p$. Likewise, from (2.24),

$$(a_1 + a_2 + \ldots + a_r)^{p^k} = a_1^{p^k} + a_2^{p^k} + \ldots + a_r^{p^k}. \qquad (2.27)$$

Putting $a_1 = a_2 = \ldots = a_r = 1$ in these equations we see that in $\mathbb{Z}_p$

$$r^p = r \qquad (2.28)$$

and, more generally,

$$r^{p^k} = r \qquad (2.29)$$

for any r in $\mathbb{Z}_p$ and positive integer k. Notice that (2.28) is essentially Fermat's theorem (see (A1.16) in Appendix 1) since it says that $r^p \equiv r \pmod{p}$ for any r in $\mathbb{Z}$.

⟩2.4 Simple and repeated roots

If a polynomial f has been split into linear factors it may be that some of the factors occur more than once. So by collecting together like factors we can write

$$f(x) = a(x - \alpha_1)^{k_1}(x - \alpha_2)^{k_2} \ldots (x - \alpha_t)^{k_t} \qquad (2.30)$$

where a is the leading coefficient of f and $\alpha_i \neq \alpha_j$ if $i \neq j$. For each factor $x - \alpha_i$ the exponent k_i is called the *multiplicity* of the root α_i. If $k_i = 1$ then α_i is a *simple* root of f and if $k_i > 1$ then α_i is a *repeated* or *multiple* root. Example 2.6 shows an extreme case when there is only one root and it is repeated as many times as the degree of the polynomial. It is just as possible to have a polynomial such as $x^3 - 1$ in $\mathbb{Q}[x]$ that has distinct linear factors in its splitting field:

$$x^3 - 1 = (x - 1)(x + \tfrac{1}{2} + i\tfrac{\sqrt{3}}{2})(x + \tfrac{1}{2} - i\tfrac{\sqrt{3}}{2}).$$

Other polynomials can have some roots that are simple and some that are repeated, as for instance the polynomial $3x^5 + 5x^4 - 40x^3 - 20x^2 + 160x - 112$ in $\mathbb{Q}[x]$ where

$$3x^5+5x^4-40x^3-20x^2+160x-112 = 3(x-\tfrac{7}{3})(x+1+\sqrt{5})^2(x+1-\sqrt{5})^2.$$

At present, if a polynomial has its coefficients in a field F, there is no way of telling whether it has any repeated roots in a splitting field unless we happen to know all the roots explicitly. It turns out that there is a straightforward procedure to test for repeated roots, and it works entirely in $F[x]$ without using any knowledge about splitting fields. To see this we shall need the idea of the *formal derivative* of a polynomial. In calculus courses we define the derivative f', or Df, of a real-valued function f by using limits and then prove that for a polynomial

$$f(x) = a_r x^r + a_{r-1} x^{r-1} + \ldots + a_1 x + a_0 \qquad (2.31)$$

we have

$$Df(x) = r a_r x^{r-1} + (r-1) a_{r-1} x^{r-2} + \ldots + a_1. \qquad (2.32)$$

In a field such as $\mathbb{Z}_p[x]$ we have no notion of a limit, but there is nothing to stop us just writing down the polynomial (2.32) and calling it the formal derivative of the polynomial $f(x)$ in (2.31). We can make this definition in any field, including $\mathbb{Q}$, and it is straightforward to check that

$$D(af) = aDf \qquad (2.33)$$

for any constant a in the field; and if $g(x) = b_s x^s + b_{s-1} x^{s-1} + \ldots + b_0$ is another polynomial we have

$$D(f + g) = Df + Dg. \qquad (2.34)$$

It is a little harder to find the derivative of the product fg; but we know that the coefficient c_k of x^k in fg is $\sum\limits_{i=0}^{k} a_i b_{k-i}$. Also the coefficient of x^k in fDg is $\sum\limits_{i=0}^{k} a_i(k-i+1)b_{k-i+1}$ and the coefficient of x^k in gDf is $\sum\limits_{j=0}^{k} b_j(k-j+1)a_{k-j+1} = \sum\limits_{i=1}^{k+1} i a_i b_{k-i+1}$. The coefficient of x^k in $fDg + gDf$ is therefore $\sum\limits_{i=0}^{k+1}(k+1)a_i b_{k-i+1} = (k+1)c_{k+1}$, and this is exactly the coefficient of x^k in $D(fg)$. So

$$D(fg) = fDg + gDf. \tag{2.35}$$

When $f = g$, (2.35) implies $D(f^2) = 2fDf$; and it is not hard to generalize that by induction to give

$$D(f^n) = nf^{n-1}Df. \tag{2.36}$$

Properties (2.35) and (2.36) can now be used to show that if f has a repeated root α (or that $x - \alpha$ is a repeated factor of f) then Df shares the same root. Suppose that

$$f(x) = (x - \alpha)^k g(x).$$

where $x - \alpha$ does not divide $g(x)$, so that $g(\alpha) \neq 0$. Then

$$\begin{aligned} Df(x) &= D[(x-\alpha)^k]g(x) + (x-\alpha)^k Dg(x) \\ &= k(x-\alpha)^{k-1}D(x-\alpha)g(x) + (x-\alpha)^k Dg(x) \\ &= k(x-\alpha)^{k-1}g(x) + (x-\alpha)^k Dg(x). \end{aligned} \tag{2.37}$$

If α is a repeated root, then $k > 1$ and it follows that $Df(\alpha) = 0$. If α is not a repeated root then $k = 1$ and $Df(x) = g(x) + (x - \alpha)Df(x)$, giving $Df(\alpha) = g(\alpha) \neq 0$.

So to test whether $f(x) \in F[x]$ has repeated roots in a splitting field we have to ask whether f and Df have a common root in an extension field of F. We can answer this question without knowing anything about the extensions of F because, as we shall see, the only way two polynomials in $F[x]$ (not necessarily a polynomial and its derivative) can share a common root in an extension field is for them to have a common non-constant factor in $F[x]$. Suppose first that the polynomials f and g in $F[x]$ have a common polynomial divisor h in $F[x]$ with $\deg(h) > 0$. Then, whether h is composite or irreducible, we know that there will be an extension of F containing a root α of h and this α will be a common root of f and g. On the other hand, if f and g have only constant factors in common, there will be polynomials $u(x)$ and $v(x)$ in $F[x]$ such that $1 = u(x)f(x) + v(x)g(x)$ and then a root β with $f(\beta) = g(\beta) = 0$ would lead to the contradiction $1 = u(\beta)f(\beta) + v(\beta)g(\beta) = 0$.

Example 2.7

Consider again the three polynomials whose roots we examined in Example 2.6 and at the beginning of this section. The polynomial $x^p - 1$ in $\mathbb{Z}_p[x]$ has derivative $px^{p-1} = 0$, so that $\mathrm{GCD}(x^p - 1, 0) = x^p - 1$. We therefore know immediately that at least some of the roots are going to be repeated and we saw in Example 2.6 that the single root is actually repeated p times. In $\mathbb{Q}[x]$ the polynomial $x^3 - 1$ has derivative $3x^2$ which clearly only has the root 0, and 0 is not a root of $x^3 - 1$. So here the polynomial and its derivative have no roots in common and we can say that $x^3 - 1$ will be a product of distinct linear factors without actually finding them. Also in $\mathbb{Q}[x]$ the polynomial $3x^5 + 5x^4 - 40x^3 - 20x^2 + 160x - 112$ has derivative $15x^4 + 20x^3 - 120x^2 - 40x + 160$. By entering these two as f and g in the computer's rational polynomial arithmetic option and then entering 'f, g' we find that they have a GCD of $x^2 + 2x - 4$ and so f will definitely have some repeated linear factors in an extension of $\mathbb{Q}[x]$.

Example 2.8

Some of the polynomials in Example 2.7 had repeated roots, but they were initially composite. No irreducible polynomials in either $\mathbb{Q}[x]$ or $\mathbb{Z}_p[x]$ have repeated roots. To see this, suppose if possible that $f(x)$ is an irreducible polynomial in $F[x]$, where F is either $\mathbb{Q}$ or $\mathbb{Z}_p$, and that f has a repeated root in an extension of F (which may be a splitting field of f). Then f and Df must have a common non-constant factor in $F[x]$; and such a factor would have to be a multiple of f (since f is irreducible). This implies that f divides Df and, because $\deg(Df) < \deg(f)$, the only way that could happen would be if $Df = 0$. In $\mathbb{Q}[x]$ we cannot have $Df = 0$ for an irreducible f, as $\deg(f) > 0$. In $\mathbb{Z}_p[x]$ we could have $Df = 0$, but only if f were of the form $a_0 + a_1 x^p + a_2 x^{2p} + \ldots + a_n x^{np}$. However, no such f is irreducible since (2.26) and (2.28) imply that

$$a_0 + a_1 x^p + a_2 x^{2p} + \ldots + a_n x^{np} = a_0^p + a_1^p x^p + a_2^p x^{2p} + \ldots + a_n^p x^{np}$$
$$= (a_0 + a_1 x + a_2 x^2 + \ldots + a_n x^n)^p.$$

$\rangle$2.5 The splitting field of $x^{p^n} - x$ in $\mathbb{Z}_p[x]$

As an extended example of some of the previous ideas let us consider the splitting field of the polynomial $x^{p^n} - x$ in $\mathbb{Z}_p[x]$. Its derivative is $p^n x^{p^n - 1} - 1 = -1$ which is relatively prime to the original polynomial so that none of its roots are repeated. The polynomial is composite in $\mathbb{Z}_p[x]$, and indeed, from (2.29), every element of $\mathbb{Z}_p$ is a root of $x^{p^n} - x$. The splitting field L will therefore be a field containing $\mathbb{Z}_p$ and containing p^n different roots $\alpha_1, \alpha_2, \ldots, \alpha_{p^n}$, so that in $L[x]$

$$x^{p^n} - x = (x - \alpha_1)(x - \alpha_2) \ldots (x - \alpha_{p^n}). \tag{2.38}$$

Of these roots, p will be the elements $0, 1, \ldots, p - 1$ of $\mathbb{Z}_p$, and $p^n - p$ will lie outside $\mathbb{Z}_p$. There can be no other roots in L, by the general result at the beginning of section 2.2 about the maximum number of roots of a polynomial; or here simply because if β were a root in L then, from (2.38), we would have

$$0 = \beta^{p^n} - \beta = (\beta - \alpha_1)(\beta - \alpha_2) \ldots (\beta - \alpha_{p^n})$$

and one of the factors $\beta - \alpha_i$ would have to be zero, giving $\beta = \alpha_i$.

The p^n roots $\alpha_1, \alpha_2, \ldots, \alpha_{p^n}$ form a field themselves with the operations of addition and multiplication in L. Certainly the set of roots is closed with respect to the operation of addition since, from the special case of the binomial theorem in any extension of $\mathbb{Z}_p$,

$$(\alpha_i + \alpha_j)^{p^n} = \alpha_i^{p^n} + \alpha_j^{p^n} = \alpha_i + \alpha_j$$

which means that $\alpha_i + \alpha_j$ is one of the roots.   Also $(-\alpha_i)^{p^n} = (-1)^{p^n} \alpha_i^{p^n} = -\alpha_i^{p^n} = -\alpha_i$ so $-\alpha_i$ is a root. Closure with respect to multiplication and taking inverses is also easy since

$$(\alpha_i \alpha_j)^{p^n} = \alpha_i^{p^n} \alpha_j^{p^n} = \alpha_i \alpha_j$$

and $(1/\alpha_i)^{p^n} = 1/(\alpha_i)^{p^n} = 1/\alpha_i$. The commutative and associative properties of addition and multiplication hold for the set of roots as they do for all elements of L. The same is true for the distributivity of multiplication over addition, and lastly the additive and multiplicative identities 0 and 1 are in the set of roots. So the set $\{\alpha_1, \alpha_2, \ldots, \alpha_{p^n}\}$ is a subfield of L. Actually the splitting field L *must be this subfield*. This is just because of the definition of the splitting field as the smallest extension field of $\mathbb{Z}_p$ which contains all the roots of $x^{p^n} - x$, and the observation that no such extension field could be smaller than the field consisting of all the roots. Just as importantly, we have shown that

for each p and n there is a field containing exactly p^n elements (2.39)

It can be shown that two finite fields with the same number of elements must be isomorphic (see Allenby(1991) or Lidl and Niederreiter(1986)), which means that the splitting field of the polynomial $x^{p^n} - x$ over $\mathbb{Z}_p$ is essentially the only field with p^n elements. Also, in the last section of Appendix 1, we show that the number of elements in any finite field must be a power of a prime, so every finite field can be thought of as a splitting field of $x^{p^n} - x$ for a suitable p and n. Towards the end of section 2.1 we saw that if there was an irreducible polynomial of degree n in $\mathbb{Z}_p[x]$ then an extension field of $\mathbb{Z}_p$ containing p^n elements could be constructed by considering $\mathbb{Z}_p[x]$ modulo that polynomial. This method can be used to produce any finite field because there are irreducible polynomials of each degree in $\mathbb{Z}_p[x]$. We shall not prove this (see Lidl and Niederreiter(1986)); but it is not hard to see that

any irreducible polynomial of degree n in $\mathbb{Z}_p[x]$

must be a factor of $x^{p^n} - x$ (2.40)

so in principle they can all be found by factorizing $x^{p^n} - x$ in $\mathbb{Z}_p[x]$. In order to prove this suppose that f is an irreducible polynomial of degree n in $\mathbb{Z}_p[x]$ and let α be a root of f in some extension field of $\mathbb{Z}_p$. Then $\mathbb{Z}_p(\alpha)$ contains p^n elements (see the paragraph just before section 2.2), and from (A1.14) in Appendix 1 each of them will satisfy $x^{p^n} = x$. In particular, α will be a root of both f and of $x^{p^n} - x$. From section 2.4 this means that f and $x^{p^n} - x$ have to share a common non-constant factor in $\mathbb{Z}_p[x]$. Since f is irreducible the only possible factors are constant multiples of itself and that gives the result.

Activity 2.9 Use the computer to factorize $x^{2^2} - x$ in $\mathbb{Z}_2[x]$ and thus verify that there is only one irreducible quadratic polynomial in $\mathbb{Z}_2[x]$ (as you already found in Activity 1.12).

Activity 2.10 Find the irreducible quadratic polynomials in $\mathbb{Z}_3[x]$ by factorizing $x^{3^2} - x$ in $\mathbb{Z}_3[x]$.

Activity 2.11 For each $n > 1$ use Eisenstein's Criterion to write down an irreducible polynomial of degree n in $\mathbb{Q}[x]$.

⟩ SUMMARY OF CHAPTER 2

At the start of chapter 2 we saw that if F is a field, such as $\mathbb{Q}$ or $\mathbb{Z}_p$, and f is an irreducible polynomial in $F[x]$ then we can use f to construct a new field that contains F as a subfield. This new field consists of all the polynomials in $F[x]$ of smaller degree than f and is denoted by $F[x]/\langle f(x)\rangle$. We form sums and products in it by taking the remainders, modulo f, of sums and products in $F[x]$ and the computer will help to do the arithmetic. This extension field of F always contains a root of f in the sense that if $K = F[x]/\langle f(x)\rangle$ then x is a root of $f(y)$ in $K[y]$. It then follows that any polynomial will have roots, either in the original field of its coefficients or in some extension field. However, as in section 2.2, if a polynomial is of degree n it can never have more than n roots in any field and may have less. The smallest extension field of F in which a polynomial $f \in F[x]$ of degree n has its full quota of n roots is called the splitting field of f with respect to F. If L is this splitting field then in $L[x]$ the polynomial $f(x)$ splits into n linear factors and it may be that some of them are repeated. Section 2.4 showed that there is a simple test to see whether f has repeated roots in an extension field and it works entirely within $F[x]$. It uses the formal derivative Df of f; and f has repeated roots if and only if f and Df have a common root in an extension field. Having a common root in an extension of F is equivalent to having a common polynomial factor in $F[x]$, and that can always be settled by Euclid's algorithm if it is not immediately clear. The last section used these ideas and also the simple form of the binomial theorem in $\mathbb{Z}_p$ and $Z_p[x]$ to investigate the splitting field of $x^{p^n} - x$ in $Z_p[x]$. It consists solely of the p^n roots of $x^{p^n} - x$ and so for each p and n there is a field containing exactly p^n elements. (Appendix 1 proves that conversely the number of elements in any finite field must be a power of a prime.) We also saw in section 2.5 that every irreducible polynomial of degree n in $Z_p[x]$ must appear as a factor of $x^{p^n} - x$.

⟩ EXERCISES FOR CHAPTER 2

1. If the natural number n is no bigger than p prove that a polynomial in $\mathbb{Z}_p[x]$ of degree n has n distinct roots in $\mathbb{Z}_p$ if and only if it divides $x^p - x$.

2. Suppose that f and g are in $\mathbb{Z}_p[x]$ and have the property that $f(a) = g(a)$ for every a in $\mathbb{Z}_p$. Prove that $f \equiv g \pmod{x^p - x}$.

3. Suppose that α is a square root of 2 in an extension field of $\mathbb{Z}_3$. That is, α is a root of the irreducible polynomial $x^2 - 2$ (or equivalently $x^2 + 1$) in $\mathbb{Z}_3[x]$. Then, as in section 2.1 just after Activity 2.3, the field $\mathbb{Z}_3(\alpha)$ contains nine elements, namely $0, 1, -1, \alpha, 1 + \alpha,$ $-1 + \alpha, -\alpha, 1 - \alpha, -1 - \alpha$. Which of these elements are powers of α? Is there an element β in $\mathbb{Z}_3(\alpha)$ such that every non-zero element in $\mathbb{Z}_3(\alpha)$ is a power of β?

4. In $\mathbb{Z}_2[x]$ the polynomial $x^2 + x + 1$ is irreducible. Suppose that α is a root of it in an extension field of $\mathbb{Z}_2$. Then the elements of $\mathbb{Z}_2(\alpha)$ are $0, 1, \alpha$ and $1 + \alpha$. Construct the addition and multiplication tables for this field.

5. Let F be any field and f an irreducible polynomial of degree n in $F[x]$, with α a root of f in some extension field of F. Prove that as the coefficients range over the elements of F the linear combinations $a_{n-1}\alpha^{n-1} + a_{n-2}\alpha^{n-2} + \ldots + a_1\alpha + a_0$ are all different by showing that if two of them were equal then we would have the contradiction that α would be a root of a polynomial of smaller degree than n. Now let $F = \mathbb{Q}$ and $f(x) = x^3 + x + 1$. If α is a root of f which element of $F(\alpha)$ is the inverse of $\alpha^2 + \alpha + 1$?

6. Following the definition of the formal derivative Df we can define the nth derivative $D^n f$ of a polynomial f by $D^1 f = Df$ and, for $n > 1$, $D^n f = D(D^{n-1} f)$. Now suppose that k is a natural number

and that f is either in $\mathbb{Q}[x]$ or $\mathbb{Z}_p[x]$ for some $p > k$. Use (2.37) to prove that α is a root of multiplicity k of f if and only if it is a root of multiplicity $k - 1$ of Df. Deduce that α is a root of multiplicity k if and only if $f(\alpha) = Df(\alpha) = \ldots = D^{k-1}f(\alpha) = 0$ and $D^k f(\alpha) \neq 0$.

7. Let p be a prime and suppose that for some natural number n the polynomial $x^{p^n} - x \in \mathbb{Z}_p[x]$ has an irreducible factor of degree s. If α is a root of such a factor write down the general form of an element of $\mathbb{Z}_p(\alpha)$. Use the binomial theorem in $\mathbb{Z}_p(\alpha)$ to show that all the p^s elements of $\mathbb{Z}_p(\alpha)$ are roots of $x^{p^n} - x$. This implies that $s \leqslant n$ since $x^{p^n} - x$ can have no more than p^n roots.

8. Let p be a prime and suppose if possible that n is the least natural number for which $x^{p^n} - x$ has an irreducible factor of degree s in $\mathbb{Z}_p[x]$ with $s \nmid n$. Show that any root of that factor would also be a root of $x^{p^s} - x$. Use this to derive a contradiction by showing that $x^{p^{n-s}} - x$ would then have the same irreducible factor of degree s with $s \nmid n - s$. Thus, for any k, each irreducible factor of $x^{p^k} - x$ in $\mathbb{Z}_p[x]$ has degree dividing k.

9. If $d \mid m$ prove that $p^d - 1$ divides $p^m - 1$ in $\mathbb{Z}$ and that $x^d - 1$ divides $x^m - 1$ in $\mathbb{Z}[x]$. Deduce that if $s \mid n$ then $x^{p^s-1} - 1$ divides $x^{p^n-1} - 1$ in $\mathbb{Z}[x]$ and thus that $x^{p^s} - x$ divides $x^{p^n} - x$ in $\mathbb{Z}[x]$ (and so in $\mathbb{Z}_p[x]$).

10. Use the results of the previous two questions to prove that in $\mathbb{Z}_p[x]$ we have
$$x^{p^n} - x = \prod_{s \mid n} \prod_{\substack{f \text{ irreduc} \\ \deg(f)=s}} f.$$

11. If f is a polynomial in $\mathbb{Z}_p[x]$ with $f(0) \neq 0$, prove that there must be a $k > 0$ with $f(x) \mid x^k - 1$. (Hint: consider the different powers of x modulo f.)

12. (Serret 1928) Prove that if a is not zero modulo the prime p then $x^p - x - a$ is irreducible in $\mathbb{Z}_p[x]$ (and so in $\mathbb{Z}[x]$). (Hint: suppose that $x^p - x - a$ has an irreducible factor $f(x)$ with $1 \leqslant \deg(f) = n < p$ and let α be a root of f in some extension field of $\mathbb{Z}_p$. Then α satisfies $\alpha^{p^n} - \alpha = 0$ (section 2.5). But α is also a root of $x^p - x - a$ so that $\alpha^p = \alpha + a$. Raise each side of this last

equality to the pth power to show that $\alpha^{p^2} = \alpha + 2a$. Find a similar expression for α^{p^n} in terms of α and a and deduce a contradiction.)

13. If k in $\mathbb{Z}_p$ is not equal to 1 prove that the polynomial $x^p - kx - a$ is always composite in $\mathbb{Z}_p[x]$ by showing that it will always have a linear factor.

⟩ Chapter 3

⟩ Quadratic integers in general and Gaussian integers in particular

⟩3.1 Algebraic numbers

Algebraic numbers are those real or complex numbers that are roots of polynomials with rational coefficients. Thus $\sqrt{2}$, $\frac{-1+\sqrt{-3}}{2}$, $\sqrt[3]{\frac{5}{7}}$ are all algebraic numbers since they are roots of the polynomials $x^2 - 2$, $x^2 + x + 1$, $7x^3 - 5$ respectively. Every rational number is an algebraic number since $\frac{a}{b}$ is the root of the linear polynomial $x - \frac{a}{b} \in \mathbb{Q}[x]$. The set of all algebraic numbers is a field with respect to the operations of complex addition and multiplication (see e.g. Stewart and Tall(1979)). In particular if α, β are algebraic numbers then so are $\alpha + \beta$, $\alpha - \beta$, $\alpha\beta$ and, if $\beta \neq 0$, also $\frac{\alpha}{\beta}$.

Example 3.1
As with $\sqrt{2}$, the numbers $\sqrt{7}$ and $\sqrt{11}$ are algebraic. So too is $\eta = \sqrt{7} + \sqrt{11}$. In this particular instance we can argue that $\eta^2 = 18 + 2\sqrt{77}$; so $(\eta^2 - 18)^2 = 308$ or $\eta^4 - 36\eta^2 + 16 = 0$. Therefore η is a root of the polynomial $x^4 - 36x^2 + 16$.

Suppose the algebraic number α is a root of $f(x) \in \mathbb{Q}[x]$ and f can be written as a product $pq \ldots$ of prime polynomials in $\mathbb{Q}[x]$. Then $f(\alpha) = p(\alpha)q(\alpha)\ldots = 0$ so that α must be a root of an irreducible rational polynomial, say $p(x)$. If α is also a root of a polynomial g, then p and g share a common root and so must have a common non-constant factor (see section 2.4). Such a factor could only be an associate of p, which means that p itself divides g. In particular, constant multiples of p are the only irreducible polynomials having α as a root. Dividing any one of them by its leading coefficient produces a unique irreducible polynomial $m(x) \in \mathbb{Q}[x]$ that has α as a root and has leading† coefficient 1. Just as for $p(x)$, each of the polymomials having α as a root is divisible by m, so m has least degree out of all such polynomials. It is called the *minimal polynomial of* α and if it has degree n then α is an *algebraic number of degree n*.

Activity 3.1 Use the computer to verify that each of the polynomials $x^2 - 2$, $x^2 + x + 1$, $x^3 - \frac{5}{7}$, $x^4 - 36x^2 + 16$ is irreducible in $\mathbb{Q}[x]$. They are therefore the minimal polynomials of the algebraic numbers mentioned in the first paragraph and in Example 3.1.

⟩3.2 Algebraic integers

Requiring a number to be a root of a polynomial with rational coefficients is the same as asking for it to be a root of a polynomial with integer coefficients, for if K is the GCD of the coefficients of f then Kf has integer coefficients and f and Kf have the same roots. The rational number $\frac{a}{b}$ for instance is the root of $bx - a \in \mathbb{Z}[x]$ as well as of $x - \frac{a}{b}$. So every algebraic number α is a root of some polynomial

$$f(x) = a_n x^n + a_{n-1} x^{n-1} + \ldots + a_1 x + a_0 \qquad (3.1)$$

where the coefficients $a_n, a_{n-1}, \ldots, a_0$ are all integers. If the leading coefficient of $f(x) \in \mathbb{Z}[x]$ is 1 (that is, if f is monic) then α is an *algebraic integer*. The numbers $\sqrt{2}$, $\frac{-1+\sqrt{-3}}{2}$ and $\sqrt{7} + \sqrt{11}$ for instance are algebraic integers; and every ordinary integer a is an algebraic integer since it is a root of $x - a \in \mathbb{Z}[x]$.

Certainly if the minimal polynomial of α has integer coefficients then α is an algebraic integer. Suppose on the other hand that α is an algebraic integer and that it is a root of a monic polynomial $f \in \mathbb{Z}[x]$ which is not necessarily its minimal polynomial. Say f is divisible by a monic polynomial g in $\mathbb{Q}[x]$; so $f = gh$ where $h \in \mathbb{Q}[x]$. At the end of section 1.7 in chapter 1 we saw that in such a case we must have $f = g'h'$ where g' and h' are integer polynomials which are just positive rational multiples of g and h respectively. The leading coefficient of g' is then positive and is a divisor of that of f, which is 1 here, so necessarily g' is monic. Since g and g' have the same leading coefficient and one is a

† Any polynomial with leading coefficient 1 is called *monic*

constant multiple of the other they must be equal. We have thus shown generally that if a monic polynomial with integer coefficients is divisible by a monic polynomial with rational coefficients then that rational factor actually has to have integer coefficients. In the situation where α is a root of the monic $f \in \mathbb{Z}[x]$ then the minimal polynomial of α must divide f and it follows immediately that the minimal polynomial has integer coefficents. So

> *an algebraic number is an algebraic integer*
>
> *if and only if its minimal polynomial has integer coefficients.* (3.2)

The number $\sqrt[3]{\frac{5}{7}}$ for instance is not an algebraic integer since its minimal polynomial is $x^3 - \frac{5}{7}$ (see Activity 3.1).

Example 3.2
The elements of $\mathbb{Z}$ are the only rational numbers that are algebraic integers; because $\frac{a}{b}$ has minimal polynomial $x - \frac{a}{b}$ and this only has integer coefficients if $\frac{a}{b} \in \mathbb{Z}$. (We also saw this in Exercise 1.13.) We shall refer to the elements of $\mathbb{Z}$ as *rational integers* when we need to distinguish them from other algebraic integers.

$\rangle$3.3 Quadratic numbers and quadratic integers

Algebraic numbers of degree 1 are roots of linear polynomials; in other words, rational numbers. The next simplest are those of degree 2, which are roots of irreducible quadratic polynomials. Each of these quadratic algebraic numbers will thus be an irrational number which is a root of a quadratic polynomial $Ax^2 + Bx + C$ where A, B and C are rational integers. The two roots of this are

$$\alpha_1 = \frac{-B + \sqrt{B^2 - 4AC}}{2A} \quad and \quad \alpha_2 = \frac{-B - \sqrt{B^2 - 4AC}}{2A} \quad (3.3)$$

and putting $D = B^2 - 4AC$ we have $\alpha_1 = \frac{-B}{2A} + \frac{\sqrt{D}}{2A}$ and $\alpha_2 = \frac{-B}{2A} - \frac{\sqrt{D}}{2A}$. Roots of the same equation, such as α_1, α_2 here, are called *conjugate* algebraic numbers. The integer D must not be a square, as that would make α_1 and α_2 rational or equivalently $Ax^2 + Bx + C$ composite. However it could be divisible by a square. For instance $A = C = 1$, $B = 4$ would give $D = 12 = 3 * 2^2$. In general if k is the largest positive integer such that k^2 divides into D then $D = k^2 d$ for an integer $d \neq 1$ that has no square divisors bigger than 1. So $\alpha_1 = \frac{-B}{2A} + \frac{k\sqrt{d}}{2A} \in \mathbb{Q}(\sqrt{d})$ and $\alpha_2 = \frac{-B}{2A} - \frac{k\sqrt{d}}{2A} \in \mathbb{Q}(\sqrt{d})$. Therefore $\mathbb{Q}(\alpha_1) \subseteq \mathbb{Q}(\sqrt{d})$ and $\mathbb{Q}(\alpha_2) \subseteq \mathbb{Q}(\sqrt{d})$. Conversely $\sqrt{d} = \frac{B}{k} + \frac{2A}{k}\alpha_1 = \frac{-B}{k} - \frac{2A}{k}\alpha_2$ is in both $\mathbb{Q}(\alpha_1)$ and $\mathbb{Q}(\alpha_2)$; and therefore $\mathbb{Q}(\alpha_1) = \mathbb{Q}(\alpha_2) = \mathbb{Q}(\sqrt{d})$.

Example 3.3
Each element of $\mathbb{Q}(\sqrt{d})$ is of the form $a + b\sqrt{d}$ where a and b are rational (see Example 2.3); and $a + b\sqrt{d}$ is a root of the polynomial $(x - a - b\sqrt{d})(x - a + b\sqrt{d}) = x^2 - 2ax + (a^2 - db^2) \in \mathbb{Q}[x]$. So every element of $\mathbb{Q}(\sqrt{d})$ is algebraic; of degree 2 if this polynomial is irreducible in $\mathbb{Q}[x]$, or 1 if it is composite.

Each quadratic algebraic integer will be a root of a polynomial $x^2 + Bx + C$ where B and C are rational integers; and so will be of the form $\frac{-B+\sqrt{B^2-4C}}{2}$ or $\frac{-B-\sqrt{B^2-4C}}{2}$. As before, writing $B^2 - 4C = k^2 d$, these can be expressed as

$$\frac{-B \pm k\sqrt{d}}{2} \tag{3.4}$$

where d and k are integers and d has no square divisors bigger than 1. We shall see that this last form can be written more meaningfully if we examine separately the cases of B odd and B even. First of all if B is odd then $B^2 - 4C$ will be odd, which means d and k must both be odd since they are divisors of $B^2 - 4C$. Further, the squares of the odd integers B and k will then be congruent to 1 modulo 4, which implies

$$1 \equiv B^2 - 4C = k^2 d \equiv d \pmod 4. \tag{3.5}$$

So in this case, putting $a = -B$ and $b = \pm k$, the quadratic integer $\frac{-B \pm k\sqrt{d}}{2}$ can be written as

$$\frac{a + b\sqrt{d}}{2} \text{ with } a, b \text{ both odd rational integers} \tag{3.6}$$

and, because of (3.5), this can only happen if $d \equiv 1 \pmod 4$. If B is even then $k^2 d = B^2 - 4C$ is divisible by 4; but d has no square divisors so it is not divisible by 4, and thus k^2 must be even. Then k is even and $\frac{-B \pm k\sqrt{d}}{2}$ is of the form

$$\frac{2e + 2f\sqrt{d}}{2} = e + f\sqrt{d}. \tag{3.7}$$

Example 3.4

For any d and any integers e, f, the number $e + f\sqrt{d}$ is a quadratic integer in $\mathbb{Q}(\sqrt{d})$ since it is a root of $x^2 - 2ex + (e^2 - df^2) \in \mathbb{Z}[x]$. If $d \equiv 1 \pmod 4$ there are other quadratic integers, because, when a, b in $\mathbb{Z}$ are both odd, we have $a^2 \equiv db^2 \pmod 4$ so that $\frac{a^2-db^2}{4}$ is a rational integer. This makes $\frac{a+b\sqrt{d}}{2}$ a quadratic integer in $\mathbb{Q}(\sqrt{d})$ since it is a root of the monic polynomial $x^2 - ax + \frac{a^2-db^2}{4} \in \mathbb{Z}[x]$. So the numbers $\sqrt{3}$, $2 + 7\sqrt{3}$, $1 - \sqrt{3}$ are quadratic integers in $\mathbb{Q}(\sqrt{3})$; just as $\sqrt{5}$, $2 + 7\sqrt{5}$, $1 - \sqrt{5}$ are in $\mathbb{Q}(\sqrt{5})$. However in $\mathbb{Q}(\sqrt{5})$ numbers such as $\frac{1+\sqrt{5}}{2}$, $\frac{7-3\sqrt{5}}{2}$ are also quadratic integers.

For a fixed $d \equiv 1 \pmod 4$ Example 3.4 shows that both types of quadratic integer (3.6) and (3.7) can occur. In (3.6) the coefficients a, b are each congruent to 1 modulo 2 and, if need be, the expression in (3.7) can also be put in the form $\frac{a+b\sqrt{d}}{2}$ with $a = 2e$ and $b = 2f$ each congruent to 0 modulo 2. So the quadratic integers of $\mathbb{Q}(\sqrt{d})$ can all be written as

$$\frac{a + b\sqrt{d}}{2} \text{ with } a \equiv b \pmod 2, \text{ when } d \equiv 1 \pmod 4. \qquad (3.8)$$

Otherwise only the type (3.7) can occur (by the remark after (3.6)) and the quadratic integers are of the form

$$a + b\sqrt{d} \text{ with any } a, b \text{ in } \mathbb{Z}, \text{ when } d \not\equiv 1 \pmod 4. \qquad (3.9)$$

Notice that

$$\frac{a + b\sqrt{d}}{2} = \left(\frac{a - b}{2}\right) + b\left(\frac{1 + \sqrt{d}}{2}\right) \qquad (3.10)$$

and if $a \equiv b \pmod 2$ then $\frac{a-b}{2}$ is an integer. So, when $d \equiv 1 \pmod 4$, each quadratic integer of the form (3.8) can also be written as

$$r + s\sigma \text{ where } \sigma = \left(\frac{1 + \sqrt{d}}{2}\right) \text{ and } r, s \text{ are integers.} \qquad (3.11)$$

Conversely, $r + s\left(\frac{1+\sqrt{d}}{2}\right) = \frac{(2r+s)+s\sqrt{d}}{2}$ with $2r + s \equiv s \pmod 2$ so that every number represented by (3.11) is automatically of the form (3.8).

When $d \equiv 1$ (mod 4) we can therefore represent quadratic integers by either of the forms (3.8) or (3.11) as convenient.

The above discussion leaves one remaining point to be tidied up. It is true that every quadratic integer in $\mathbb{Q}(\sqrt{d})$ must be represented either by (3.8) or (3.9). However these expressions only give numbers of degree 2 if they are irrational, which is when $b \neq 0$. When $b = 0$ they both represent ordinary rational integers, which are the algebraic integers of degree 1. Example 3.3 shows that there can be no algebraic integers of degree higher than 2 in $\mathbb{Q}(\sqrt{d})$. So *all* the algebraic integers in $\mathbb{Q}(\sqrt{d})$ must be represented by the appropriate choice of (3.8) or (3.9); or equivalently, by (3.9) or (3.11).

Notice what happens when we add, subtract, or multiply two quadratic integers in $\mathbb{Q}(\sqrt{d})$. If $d \not\equiv 1$ (mod 4) and $a_1 + b_1\sqrt{d}$, $a_2 + b_2\sqrt{d}$ are two integers of $\mathbb{Q}(\sqrt{d})$ then

$$(a_1 + b_1\sqrt{d}) \pm (a_2 + b_2\sqrt{d}) = (a_1 \pm a_2) + (b_1 \pm b_2)\sqrt{d} \qquad (3.12)$$

and

$$(a_1 + b_1\sqrt{d})(a_2 + b_2\sqrt{d}) = (a_1a_2 + db_1b_2) + (a_1b_2 + a_2b_1)\sqrt{d}. \quad (3.13)$$

So in this case the set $\mathcal{D} \subseteq \mathbb{C}$ of quadratic integers in $\mathbb{Q}(\sqrt{d})$ is closed under these operations and is thus a subring of $\mathbb{Q}(\sqrt{d})$, and of $\mathbb{C}$, since the other necessary properties are inherited from $\mathbb{C}$. Indeed it is an integral domain (see Appendix 1) since the cancellation laws hold throughout $\mathbb{C}$. If $d \equiv 1$ (mod 4) we may take two typical integers in $\mathbb{Q}(\sqrt{d})$ to be $r_1 + s_1\sigma$ and $r_2 + s_2\sigma$ where σ is as in (3.11). Here

$$(r_1 + s_1\sigma) \pm (r_2 + s_2\sigma) = (r_1 \pm r_2) + (s_1 \pm s_2)\sigma \qquad (3.14)$$

and

$$(r_1 + s_1\sigma)(r_2 + s_2\sigma) = r_1r_2 + (r_1s_2 + r_2s_1)\sigma + s_1s_2\sigma^2$$

$$= (r_1r_2 + \tfrac{d-1}{4}s_1s_2) + (r_1s_2 + r_2s_1 + s_1s_2)\sigma \qquad (3.15)$$

since $\sigma^2 = \sigma + \frac{d-1}{4}$. So again $\mathcal{D}$ is closed under these operations and, as before, is an integral domain.

⟩3.4 The integers of $\mathbb{Q}(\sqrt{-1})$

In order to gain more experience of working with quadratic integers, let us look at the integers of $\mathbb{Q}(\sqrt{-1})$. From (3.9) they are all of the form $a + b\sqrt{-1}$ where a and b are rational integers; in other words they are complex numbers with integer real and imaginary parts and make up the set $\mathbb{Z}[\sqrt{-1}]$. These particular quadratic integers are often called *Gaussian integers* after C. F. Gauss who first investigated them. We can see part of the complex plane and move about among the Gaussian integers if we choose option 2, 'Arithmetic with quadratic integers', from the main menu of the accompanying computer programs. The computer will denote $\sqrt{-1}$ by either i or j, depending on your preference; but for convenience on the printed page I shall use i here. The first screen of this option shows a square occupying most of the display and giving a view of part of the complex plane. The origin, $0 + 0i$, is at the centre of the square and a small yellow cross is also there initially. Some of the points of the complex plane are indicated by dots of different colours, and we shall see the significance of these colours later. We can move the cross over the points of the plane using the cursor keys and, each time it is moved to a Gaussian integer, a side panel shows which point it is currently covering. The side display also shows a rational integer associated with each Gaussian integer and called its *norm*. The norm of $a + bi$ is $a^2 + b^2 = (a + bi)(a - bi)$ and geometrically this represents the square of the distance from the origin to the point $a + bi$.

Activity 3.2 What are the norms of the following Gaussian integers: $3 + 7i$, $3 - 7i$, $2 + 3i$, -13, $-13i$, $17 - 13i$, $-13 - 17i$?

As well as finding norms for us the computer also makes it easy to add, subtract, multiply or divide Gaussian integers. Suppose for instance that we want to add $5 + 17i$ and $8 - 11i$. We first move the cross shaped cursor to $5 + 17i$ and press the key† marked '+'. The bottom of the screen immediately shows

$(5 + 17i)+$

and the computer waits for us to choose the next summand. We do that by moving the cross to $8 - 11i$ and pressing the ENTER (or RETURN) key. The display then becomes

$(5 + 17i) + (8 - 11i) = 13 + 6i$.

---

† On most keyboards you have to press the 'SHIFT' key at the same time as '+'.

Performing subtraction is just as easy. If we now wanted to subtract $21 + 4i$ from $13 + 6i$, we should move the cursor to $13 + 6i$ and press the key marked '$-$'. Then moving it to $21 + 4i$ and pressing ENTER would then make the display show

$(13 + 6i) - (21 + 4i) = -8 + 2i$.

Activity 3.3 Find the sums and differences of the following pairs of Gaussian integers.
(i) $21 + 12i$ and $9 - 7i$; (ii) $-11 - 4i$ and $-16 + 8i$; (iii) $2 - 26i$ and $-1 - 9i$; (iv) $12 - 8i$ and $-7 + 2i$.

Finding the product of two integers in this domain is very similar. To multiply $4 + i$ by $1 + i$ we move the cursor to $4 + i$, press '$*$', and then move the cursor to $1 + i$ and press ENTER. Of course we could have first moved the cursor to $1 + i$, pressed '$*$', and then moved to $4 + i$ followed by ENTER. The product of any two quadratic integers in the visible region is obtained in a like manner.

Activity 3.4 Find the products of the following pairs of Gaussian integers:
(i) $4 + i$ and $1 + i$; (ii) $1 + 2i$ and $1 - 2i$; (iii) $17 - 11i$ and $7i$; (iv) $3 - 7i$ and $2 + 3i$; (v) $-11 + 4i$ and $-6 - 9i$; (vi) $13 + 2i$ and $-1 + i$.

There is a very useful observation that can be made about the norms of Gaussian integers that are multiplied together. The numbers $5 - 2i$ and $3 + i$, for example, have norms 29 and 10 respectively, while their product $17 - i$ has norm $290 = 29 * 10$. The norm of a product always bears that same relationship to the original norms, because the numbers $a + bi$ and $c + di$ have norms $a^2 + b^2$ and $c^2 + d^2$ respectively, and their product is $ac - bd + (ad + bc)i$ which has norm $(ac - bd)^2 + (ad + bc)^2 = a^2c^2 + b^2d^2 + a^2d^2 + b^2c^2 = (a^2 + b^2)(c^2 + d^2)$. So

> *the norm of a product is always equal to the product*
> *of the norms of the factors.*   (3.16)

We have proved that (3.16) holds for a product of two factors. It is easy to see that it remains true for a product of $n > 2$ factors. Suppose that $\gamma_1 \gamma_2 \ldots \gamma_n$ is a product of n Gaussian integers, some of which may be equal. Then, denoting the norm by $N(\ )$ and repeatedly applying the case of (3.16) for two factors, we see that $N(\gamma_1 * \gamma_2 \ldots \gamma_n) = N(\gamma_1) * N(\gamma_2 \ldots \gamma_n) = N(\gamma_1) * N(\gamma_2) * N(\gamma_3 \ldots \gamma_n) = N(\gamma_1) * \ldots * N(\gamma_n)$.

We can also divide one Gaussian integer into another in a similar way. We first move the cursor to the number that we want to divide and press the key marked '/'. Then we move to the number that we are going to divide by and press ENTER. As in any integral domain that is not a field, division is sometimes exact and sometimes not. Thus when $7 + 23i$ is divided by $4 + i$ the result is $3 + 5i$ since $7 + 23i = (4 + i)(3 + 5i)$. However we cannot similarly divide $9 + 11i$ by $-1 + 2i$ with an exact result in $\mathbb{Z}[i]$ because $9 + 11i$ cannot be written as $(-1 + 2i)(x + yi)$ for any Gaussian integer $x + yi$. If it could, then Norm$(9+11i) = 202$ would be the product of Norm$(-1 + 2i) = 5$ and Norm$(x + yi) = x^2 + y^2 \in \mathbb{Z}$; but 202 is not 5 times any integer. In this case the best we can do is to write $9 + 11i$ as a multiple of $-1 + 2i$ plus a remainder in $\mathbb{Z}[i]$; as, for instance,

$$9 + 11i = (2 + 3i)(-1 + 2i) + 17 + 10i$$
$$= (2 - 5i)(-1 + 2i) + 1 + 2i$$
$$= (3 - 6i)(-1 + 2i) - i.$$

This last representation is the computer's choice, which it writes as

$$(9 + 11i)/(-1 + 2i) = (3 - 6i) + \frac{-i}{-1 + 2i}.$$

Activity 3.5 In each of the following cases find quotients and remainders when the first Gaussian integer is divided by the second.
(i) $-1 + 5i, i$; (ii) $5, 1 - 2i$; (iii) $7 + 4i, 3 + 2i$; (iv) $3 + 11i, -6 + 5i$;
(v) $27 - 5i, 3 - 7i$; (vi) $19 - 5i, -3 + 2i$.

$\rangle$3.5 Division with remainder in $\mathbb{Z}[i]$

In those cases where there is not exact division in $\mathbb{Z}[i]$ the computer finds a quotient and remainder such that the norm of the remainder is smaller than that of the divisor. To see that this is always possible, suppose that we try to divide $a + bi$ by $x + yi \neq 0 + 0i$. In $\mathbb{Q}(i)$ we have

$$\frac{a + bi}{x + yi} = \frac{(a + bi)(x - yi)}{(x + yi)(x - yi)} = \frac{(a + bi)(x - yi)}{x^2 + y^2}$$

so we try to divide the product $(a+bi)(x - yi) = (ax + by) + (bx - ay)i$ by the norm $N = x^2 + y^2$ of $x + yi$. Using ordinary division with remainder in $\mathbb{Z}$ we can certainly write $ax + by = N * p + R$ for some integers p and R with $|R| \leqslant \frac{N}{2}$; and likewise $bx - ay = N * q + S$ for integers q and S with $|S| \leqslant \frac{N}{2}$. Then

$$(a + bi)(x - yi) = N * (p + qi) + R + Si \qquad (3.17)$$

or equivalently

$$\frac{a + bi}{x + yi} = p + qi + \frac{R + Si}{N}. \qquad (3.18)$$

Since $N = (x - yi)(x + yi)$, (3.17) shows that $R + Si = (a + bi)(x - yi) - N * (p + qi)$ is exactly divisible by $x - yi$. If $R + Si = (x - yi)(r + si)$ for integers r and s, then from either (3.17) or (3.18)

$$(a + bi) = (p + qi) * (x + yi) + r + si \qquad (3.19)$$

and from the inequalities satisfied by R, S we see that

$$\text{Norm}(r + si) = \text{Norm}(R + Si)/\text{Norm}(x - yi) = \frac{R^2 + S^2}{N} \leqslant \frac{N^2/2}{N}$$

$$= \frac{N}{2} = \frac{1}{2}\text{Norm}(x + yi). \qquad 3.20$$

The equation (3.19), and the inequality $\text{Norm}(r + si) < \text{Norm}(x + yi)$ in (3.20), express the 'division with remainder of smaller norm' property that we wanted. As we have seen, the computer usually writes (3.19) as

$$(a + bi)/(x + yi) = p + qi + \frac{r + si}{x + yi}. \qquad (3.21)$$

The choice of the integer quotient $p + qi$ can be given a nice geometrical interpretation if we first observe that every point in the complex plane lies inside or on the boundary of a square whose vertices are Gaussian integers. The sides of the smallest such squares have length 1 so even their centres are no further from their corners than $1/\sqrt{2}$. Now, from (3.18), the square of the distance between $p + qi$ and $\frac{a+bi}{x+yi}$ is $\left(\frac{R}{N}\right)^2 + \left(\frac{S}{N}\right)^2 \leqslant \frac{1}{2}$. Therefore $p + qi$ is within $1/\sqrt{2}$ of $\frac{a+bi}{x+yi}$ and is thus one of the four Gaussian integers forming the corners of the square of side 1 containing $\frac{a+bi}{x+yi}$ (see Figure 3.1 for the case of $\frac{a+bi}{x+yi} = \frac{-11-8i}{3+4i}$). Sometimes there is only one corner whose distance from $\frac{a+bi}{x+yi}$ is less than $1/\sqrt{2}$, as when $a + bi = 9 - 37i$, $x + yi = 7 + 3i$ (see Activity 3.6 below). Sometimes

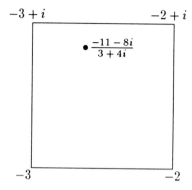

Figure 3.1   The four Gaussian integers surrounding $\frac{-11-8i}{3+4i}$.

there are two Gaussian integers within $1/\sqrt{2}$ of $\frac{a+bi}{x+yi}$, as is the case with $a+bi = -11-8i$, $x+yi = 3+4i$; and if $\frac{a+bi}{x+yi}$ is exactly in the centre of the unit square formed by its surrounding Gaussian integers, there are four equally possible choices for $p+qi$.

Activity 3.6 The complex number $\frac{9-37i}{7+3i}$ lies inside the unit square whose vertices are the points $-1-5i$, $-5i$, $-4i$, $-1-4i$. Use pencil and paper (and maybe some computer calculations) to show that just one of these points is within $1/\sqrt{2}$ of $\frac{9-37i}{7+3i}$. You may find it helpful to first ask the computer to divide $9-37i$ by $7+3i$. That will give $\frac{1+i}{7+3i}$ as the difference between $\frac{9-37i}{7+3i}$ and one of the corners of the square. The modulus of this difference is then $\frac{|1+i|}{|7+3i|} = \frac{\sqrt{2}}{\sqrt{7^2+3^2}} < \frac{1}{\sqrt{2}}$ and the other differences are obtained by adding either -1, $-i$ or $-1-i$ to $\frac{1+i}{7+3i}$.

Activity 3.7 The complex number $\frac{-11-8i}{3+4i}$ lies inside the unit square whose vertices are the points -3, -2, $-2+i$ and $-3+i$. Show that two of these points are within $1/\sqrt{2}$ of $\frac{-11-8i}{3+4i}$.

Activity 3.8 Show that the complex number $\frac{4+11i}{1-i}$ is equidistant from the four Gaussian integers $-4+7i$, $-3+7i$, $-3+8i$ and $-4+8i$.

When choosing $p+qi$ the computer always picks the nearest Gaussian integer to $\frac{a+bi}{x+yi}$, or one of the nearest if there is more than one. Notice that when the nearest Gaussian integer is the origin $0+0i$ we would have $p+qi = 0+0i$ and $r+si = a+bi$. This happens for instance with $a+bi = -1+2i$ and $x+yi = 9+11i$. In cases like this the right-hand side of (3.21) would be identical to the left. So, rather than just repeating the same fraction, the computer 'rationalizes' the denominator of $\frac{a+bi}{x+yi}$ by multiplying both top and bottom of the fraction by $x - yi$, obtaining $\frac{a+bi}{x+yi} = \frac{R+Si}{N}$. Thus when we ask the computer to divide $-1 + 2i$ by $9 + 11i$ it puts

$$(-1 + 2i)/(9 + 11i) = \frac{13 + 29i}{202}.$$

Activity 3.9 In each part of Activity 3.5 divide the second Gaussian integer by the first.

The fact that we can always perform division either exactly or with a remainder of smaller norm, together with the multiplicative nature of the norm as in (3.16), means that this norm is Euclidean and the Gaussian integers form a Euclidean domain (see Appendix 1). So every two Gaussian integers have a greatest common divisor which can be found using Euclid's idea of systematic division with remainder just as we used Euclid's algorithm to find greatest common divisors of polynomials in section 1.4. To find a GCD of $17 + 25i$ and $7 + 9i$ for instance, you first move the cross on the screen to $17 + 25i$ and press '/'. Then move to $7 + 9i$ and press ENTER. This produces

$$(17 + 25i)/(7 + 9i) = 3 + \frac{-4 - 2i}{7 + 9i}. \tag{3.22}$$

Press '/' immediately after this and move the cross to the remainder $-4 - 2i$. Pressing ENTER now gives

$$(7 + 9i)/(-4 - 2i) = -2 - i + \frac{1 + i}{-4 - 2i}. \tag{3.23}$$

Again press '/' and move the cross to the last remainder, namely $1 + i$. This time when you press ENTER you get

$$(-4 - 2i)/(1 + i) = -3 + i. \tag{3.24}$$

Equation (3.22), which could equally be written as $(17 + 25i) = 3(7 + 9i) + (-4 - 2i)$, implies that any common divisor of $17 + 25i$ and $7 + 9i$ has to divide $-4 - 2i$; and on the other hand any common divisor of $7 + 9i$ and $-4 - 2i$ has to divide $17 + 25i$. So any greatest common divisor of $17 + 25i$ and $7 + 9i$ must also be a greatest common divisor of $7 + 9i$ and $-4 - 2i$. In the same way equation (3.23) implies that each GCD of $7 + 9i$ and $-4 - 2i$ is a GCD of $-4 - 2i$ and $1 + i$. Finally equation (3.24) says that $1 + i$ is itself a GCD of $-4 - 2i$ and $1 + i$ since it is a divisor of $-4 - 2i$. So $1 + i$ is a GCD of $17 + 25i$ and $7 + 9i$. If you initially decided to divide $7 + 9i$ by $17 + 25i$, then instead of (3.22) you would obtain

$$(7 + 9i)/(17 + 25i) = \frac{344 - 22i}{914}.$$

Here you should press '/' next and move the cross back to $7+9i$. Pressing ENTER then gives (3.22) and the remaining steps can be continued as before. Once started, we keep on dividing the latest remainder into the current divisor and the first number that divides exactly is always a GCD. The entire procedure can be carried out very easily since the steps are the same at every stage.

Activity 3.10 Find greatest common divisors of the following pairs of Gaussian integers.
(i) $1 + 29i$ and $-21 + 22i$; (ii) $14 + 6i$ and 16; (iii) $13 - 2i$ and $6 - 5i$; (iv) $-5 + 12i$ and $8 + 8i$; (v)$-3 + 15i$ and $9 + 12i$.

Once we have found one GCD γ of two numbers we can write down all of them. This is because any other GCD γ' must be a unit multiple of γ, as shown in section 3 of appendix 1 for any integral domain. In $\mathbb{Z}[i]$ the elements $1, -1, i$ and $-i$ are invertible, and if $\upsilon \in \mathbb{Z}[i]$ is a unit with inverse υ' then (again using $N(\ )$ for the norm) we have $N(\upsilon)N(\upsilon') = N(\upsilon\upsilon') = N(1) = 1$, so the rational integer $N(\upsilon)$, being a divisor of 1, must be ± 1. This leads to $\upsilon = \pm 1$ or $\pm i$. So $1, -1, i$ and $-i$ are the only unit Gaussian integers and on the computer screen they appear in white.

Activity 3.11 Write down all the greatest common divisors of the pairs in Activity 3.10.

⟩3.6 Prime and composite integers in $\mathbb{Z}[i]$

As we saw with polynomials in chapter 1, we can classify the non-zero and non-unit elements into *irreducible* and non-irreducible or *composite* elements. An irreducible Gaussian integer α has no Gaussian integer factors apart from units and unit multiples (or *associates*) of α, and a composite Gaussian integer can be written as a product of two factors neither of which is a unit.

Example 3.5
Suppose that the number $3 + 2i$ can be written as $\gamma_1 \gamma_2$ for some Gaussian integers γ_1 and γ_2. Then $\text{Norm}(\gamma_1)\text{Norm}(\gamma_2) = \text{Norm}(3 + 2i) = 13$. So one of the positive whole numbers $\text{Norm}(\gamma_1)$, $\text{Norm}(\gamma_2)$ must be 1 and the other must be 13. The factor that has norm 1 must then be ± 1 or $\pm i$, so $3 + 2i$ can only be written as a product of Gaussian integers if one of the factors is a unit. In other words $3 + 2i$ is irreducible in $\mathbb{Z}[i]$.

The argument of this example can be applied to any Gaussian integer whose norm is prime in $\mathbb{Z}$. If such a Gaussian integer could be written as a product of two factors it would follow that one of the factors had norm 1 and thus was a unit. So a Gaussian integer with a norm that is prime in $\mathbb{Z}$ has to be irreducible in $\mathbb{Z}[i]$.

Activity 3.12 Use the computer to see how many Gaussian integers $a + bi$ you can find with $|a| < 5$, $|b| < 5$, and whose norm $a^2 + b^2$ is prime in $\mathbb{Z}$. (There are 28 altogether.)

Activity 3.13 On the computer screen the irreducible elements are shown in red. Can you find any irreducible Gaussian integers whose norms are not prime in $\mathbb{Z}$?

In $\mathbb{Z}[i]$ the irreducible elements play a basic role (just as they do in $\mathbb{Q}[x]$ and in $\mathbb{Z}$) since every non-zero integer in $\mathbb{Z}[i]$ is either a unit, is irreducible, or is a product of irreducible integers in $\mathbb{Z}[i]$. In chapter 1 we proved the corresponding statement for polynomials by using induction on their degrees. Here we use induction on the norms of Gaussian integers. As we have seen, the Gaussian integers of norm 1 are units. Those of norm 2 must be irreducible since 2 is prime in $\mathbb{Z}$ (see Example 3.5 and following remarks). Suppose then that we have reached the stage of having proved that for some rational integer $n \geqslant 2$ every non-zero Gaussian integer with norm not exceeding n is either a unit, is irreducible or a product of irreducibles. Consider the next possible norm $n + 1$. If there are no composite numbers (or no numbers at all!) with norm $n + 1$ we have proved what we want for norms not exceeding $n + 1$ and we replace the bound n by $n + 1$. Otherwise let α be a composite Gaussian integer of norm $n + 1$. Then, by definition, α will be the product of two integers β and γ, neither of which is a unit, and $\alpha = \beta\gamma$ implies

$N(\alpha) = N(\beta)N(\gamma)$. Thus

$$N(\beta)N(\gamma) = n + 1 \qquad (3.25)$$

and $N(\beta) > 1$, $N(\gamma) > 1$ since they are not units. So (3.25) gives $1 < N(\beta) \leqslant n$ and $1 < N(\gamma) \leqslant n$, whence, by assumption, β and γ are each either irreducible or can be written as products of irreducible elements. Writing $\beta = \beta_1 \ldots \beta_r$ and $\gamma = \gamma_1 \ldots \gamma_s$, (where $\beta_1, \ldots, \beta_r, \gamma_1, \ldots, \gamma_s$ are each irreducible and $r = 1$ or $s = 1$ if β or γ is irreducible) we have $\alpha = \beta\gamma = \beta_1 \ldots \beta_r\gamma_1 \ldots \gamma_s$. So we can always make the inductive step from norms not exceeding n to norms not exceeding $n + 1$ and the result is therefore true generally.

In a Euclidean domain every irreducible element p has the prime property: if p divides a product it must divide one of the factors. This was shown in chapter 1, section 6 for polynomial domains and in section 5 of appendix 1 it is proved for any Euclidean domain. So in the particular Euclidean domain $\mathbb{Z}[i]$ every irreducible Gaussian integer is prime. Most Gaussian primes have norms that are rational primes, but Activity 3.13 shows that there are exceptions. Later in chapter 5 we shall see why this is so.

Activity 3.14 How many Gaussian primes are there (i) with norm 2; (ii) with norm 5?

Example 3.6
If $a + bi$ had norm $a^2 + b^2 = 3$ then $a^2 \leqslant 3$ would give $a = -1$, 0 or 1. Similarly b would have to be -1, 0 or 1. Then $a^2 + b^2$ would be at most 2. So there is no Gaussian integer whose norm is 3.

Activity 3.15 Find bounds, as in the last example, for the real and imaginary parts of a Gaussian integer with norm 7. Then use the computer to investigate the resulting possibilities and thus show that there is no Gaussian integer whose norm is 7.

Naturally we can only see a limited number of Gaussian primes on the computer screen, but there are in fact an infinite number of them. This can be proved using essentially the same idea as in Euclid's proof that there are an infinite number of primes in $\mathbb{Z}$. Suppose that $\pi_1, \pi_2, \ldots, \pi_n$ is any finite list of Gaussian primes. We form their product Π which will be a non-zero integer in $\mathbb{Z}[i]$ and then add a unit υ to Π. We want to avoid any possibility of $\Pi + \upsilon$ being zero or a unit, and a simple way to ensure that it is neither of these is to choose $\upsilon = 1$ if the real part of

Π is at least 0, and $\upsilon = -1$ otherwise. Then $\Pi + \upsilon$ is a non-zero and non-unit element in $\mathbb{Z}[i]$ so it is either prime or a product of primes. We let π be the name of a prime dividing $\Pi + \upsilon$ and note that π cannot be equal to any of the primes $\pi_1, \pi_2, \ldots, \pi_n$ in our initial list nor to any of their associates. If for instance π were to divide π_i then it would divide $\Pi = \pi_1 \pi_2 \ldots \pi_n$ and also $\Pi + \upsilon$ (by assumption), which would mean that π had to divide υ; but that cannot happen as no prime divides a unit (see Example 3.7 below). So π is different from each of our previous primes. This means that the set of Gaussian primes cannot be contained in any finite list and that is the required result.

Example 3.7
Suppose υ is a unit with inverse υ' and that α is an element that divides υ. Then α divides $\upsilon * \upsilon' = 1$ and so must itself be a unit. Therefore no non-unit element can divide a unit.

As well as infinitely many primes there are infinitely many composite Gaussian integers such as $2(1 + i), 3(1 + i), \ldots,$ or $k(1 + i)$ for any non-unit k. This will also follow immediately when we prove an interesting geometrical result: there are arbitrarily large squares in the complex plane that only contain composite Gaussian integers. In the case of $\mathbb{Z}$ the corresponding statement is that there are arbitrarily long intervals containing only composite integers, and we prove that by taking any rational integer $k \geqslant 2$ and looking at the consecutive rational integers $k!+2, k!+3, \ldots, k!+k$. Each of these numbers is composite since the first is bigger than 2 and is divisible by 2, the second (if $k \geqslant 3$) is bigger than 3 and is divisible by 3, $\ldots$, and the last is bigger than k and is divisible by k. In $\mathbb{Z}[i]$ we construct a product analogous to a factorial in $\mathbb{Z}$ by choosing some natural number k and multipling together all the Gaussian integers $a + bi$ with $0 < a \leqslant k$ and $0 < b \leqslant k$. This produces a Gaussian integer Γ, say, and we then look at each of the integers $\Gamma + a + bi$ with $0 < a \leqslant k$ and $0 < b \leqslant k$. They all turn out to be composite. If $k = 1$ the only allowable number $a + bi$ is $1 + i$. So in this case Γ is $1 + i$ and $\Gamma + 1 + i = 2 + 2i$ which is composite. In general, no possible number $a + bi$ is a unit and for each one the number Γ, and so $\Gamma + a + bi$, is divisible by $a + bi$. Also when $k \geqslant 2$, the product Γ includes the numbers $1 + i,\ 1 + 2i,\ 2 + i$ and $2 + 2i$ whose moduli are 2, 5, 5 and 8 respectively. So, for any $a + bi$ and any $k \geqslant 2$ we have $|\Gamma| > 2*5*|a + bi|$, whence $|\Gamma + a + bi| \geqslant |\Gamma| - |a + bi| > 9*|a + bi|$. Therefore $\Gamma + a + bi$ cannot be an associate of $a + bi$ and so is composite (being the product of the non-unit $a + bi$ and another factor which is a non-unit as its modulus is more than 9). The square in the complex plane whose corners are the Gaussian integers $\Gamma + 1 + i$, $\Gamma + k + i$, $\Gamma + k + ki$ and $\Gamma + 1 + ki$, and whose sides are $k - 1$ units long, then contains only composite elements of $\mathbb{Z}[i]$. On the computer screen it is always easy to recognise composite integers since they are either all blue or all black (depending on your choice).

Example 3.8
When $k = 2$ we form the product of the four Gaussian integers $1+i$, $1+2i$, $2 + i$ and $2 + 2i$. This gives $\Gamma = -20$ and the four composite numbers constructed as above are $-20+(1+i) = -19+i$, $-20+(2+i) = -18+i$, $-20 + (2 + 2i) = -18 + 2i$ and $-20 + (1 + 2i) = -19 + 2i$.

Activity 3.16 Use the computer to verify that the four numbers constructed in Example 3.8 are composite by showing that $-19 + i$ is divisible by $1 + i$, $-18 + i$ by $2 + i$, $-18 + 2i$ by $2 + 2i$ and $-19 + 2i$ by $1 + 2i$.

Activity 3.17 Try to find four composite numbers that form the corners of a 1×1 square and that are nearer to the origin than the numbers $-19+i$, $-18 + i$, $-18 + 2i$ and $-19 + 2i$.

If we followed the above construction to try to find nine composite integers in $\mathbb{Z}[i]$ which form a square we should first have to multiply the nine numbers $1 + i$, $2 + i$, $3 + i$, $1 + 2i$, $2 + 2i$, $3 + 2i$, $1 + 3i$, $2 + 3i$, $3 + 3i$. Their product is $\Gamma = 7800 + 7800i$ and the required square would have corners $7800 + 7800i + 1 + i = 7801 + 7801i$, $7800+7800i+3+i = 7803+7801i$, $7800+7800i+3+3i = 7803+7803i$ and $7800 + 7800i + 1 + 3i = 7801 + 7803i$. The computer will not actually do all the required multiplication for us, nor show the resulting square, as $7800 + 7800i$ lies very far from the visible region on the screen. However there are examples much nearer the origin. The numbers $10-12i$, $11-12i$, $12-12i$, $10-11i$, $11-11i$, $12-11i$, $10-10i$, $11-10i$ and $12 - 10i$ for instance form a 2×2 square and are all composite. If you have a high resolution display (such as EGA or VGA) and ask the program to change the scale of the picture you should also be able to find examples of larger squares containing only composite integers. Try first looking at the region around $-56 + 18i$ or around $-6 - 62i$, and then at squares whose lower left-hand corner is $-78 - 70i$.

You may already have noticed that the pattern of prime and composite integers in $\mathbb{Z}[i]$ is highly symmetric. If $a + bi$ is composite and υ is a unit with inverse υ' then $\upsilon * (a + bi)$ is also composite and if $\upsilon * (a + bi)$

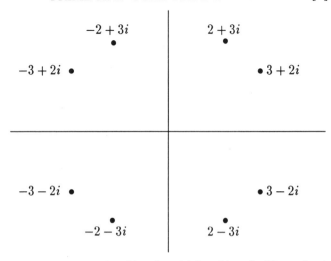

Figure 3.2   The prime $3 + 2i$ with unit multiples of it and of its conjugate $3 - 2i$.

is known to be composite then so is $a + bi = \upsilon' * \upsilon * (a + bi)$. So if one of $a + bi$, $i * (a + bi) = -b + ai$, $-1 * (a + bi) = -a - bi$ or $-i * (a + bi) = b - ai$ is composite then they all are, and conversely if any are prime then they all are. Similarly if either α or its complex conjugate $\bar{\alpha}$ is composite then so is the other since $\alpha = \beta\gamma$ is equivalent to $\bar{\alpha} = \overline{\beta\gamma} = \bar{\beta}\bar{\gamma}$. So again the prime or composite character of $a + bi$ is the same as that of each of $\overline{a + bi} = a - bi$, $i * (a - bi) = b + ai$, $-1 * (a - bi) = -a + bi$ and $-i * (a - bi) = -b - ai$. In other words the Gaussian integers

$$a + bi, -b + ai, -a - bi, b - ai, a - bi, b + ai, -a + bi, -b - ai \quad (3.26)$$

are either all prime or all composite. They can be obtained by starting with $a + bi$, rotating it round the origin by $90°$, $180°$, $270°$, and taking conjugates of the resulting numbers. If a and b are not equal and are both non-zero then these eight numbers are all different and are distributed round the origin as illustrated in Figure 3.2 for $a = 3$, $b = 2$. If either a or b is zero and the other non-zero, or if a and b are equal and non-zero, there are only four distinct numbers in the set (3.26). For instance when $a = 2$, $b = 0$, the only numbers mentioned in (3.26) are $2, 2i, -2, -2i$; and when $a = b = 3$, those listed are $3 + 3i, -3 + 3i, -3 - 3i, 3 - 3i$. Taking conjugates in these cases does not produce any new points.

Example 3.9
The integers $3 + 3i$, $4 + 3i$, $4 + 4i$, $3 + 4i$ form a small square and are all composite. By (3.26) so are the numbers $-3 + 3i$, $-3 + 4i$, $-4 + 4i$, $-4 + 3i$; the numbers $-3 - 3i$, $-4 - 3i$, $-4 - 4i$, $-3 - 4i$; and the numbers $3 - 3i$, $3 - 4i$, $4 - 4i$, $4 - 3i$; and each of these sets forms a square.

Activity 3.18 The four numbers $-19 + i$, $-18 + i$, $-18 + 2i$ and $-19 + 2i$ are composite (see Example 3.8 and Activity 3.16). Use (3.26) with $a = -19$, $b = 1$ to write down the other seven numbers derived from $-19 + i$ by successive 90° rotations and conjugation. Identify them on the computer screen and verify that they form 1×1 squares of composite numbers with the numbers derived similarly from $-18 + i$, $-18 + 2i$ and $-19 + 2i$.

Activity 3.19 We saw above that $10 - 12i$ is a corner of a 2×2 square of composite numbers. Use (3.26) to find three other 2×2 squares of composite numbers derived from this one. When $a = 10$ and $b = -12$, (3.26) gives eight distinct points which should all be corners of squares of composite numbers. Does this mean that you can use (3.26) to find eight different squares of composite numbers derived from the one containing $10 - 12i$?

Just as any composite polynomial has an essentially unique expression as a product of irreducible factors (see section 1.6), so two expressions for a composite Gaussian integer as a product of primes must have the same number of factors, and each prime occurring in one of the expressions must be a unit multiple of some prime in the other. For example $19 + 13i$ can be written as

$$(1 + 2i)(1 + i)(2 - 7i)$$

or

$$(1 + i)(2 - i)(7 + 2i)$$

or

$$(-2 + 7i)(i - 1)(i - 2).$$

Each factor mentioned is prime and, comparing for instance the factors in the second product with those in the first, we see that $2 - i = -i * (1 + 2i)$ and $7 + 2i = i * (2 - 7i)$.

Activity 3.20 Show that each factor in the product $(-2 + 7i)(i - 1)(i - 2)$ is a unit multiple of one of the factors in the product $(1 + 2i)(1 + i)(2 - 7i)$.

(You could use the computer to divide $-2 + 7i$ by each of $1 + 2i$, $1 + i$, $2 - 7i$ in turn and check that there is exact division with a unit quotient in one instance. Then do the same with $i - 1$ and $i - 2$.)
In any Euclidean domain there is unique factorization of composite elements into irreducible factors as above and the general proof is given in Appendix 1.

The computer will actually factorize composite Gaussian integers automatically for us. To factorize $10 - 11i$ for instance, we move the yellow cross to $10 - 11i$ and just press ENTER. The display immediately shows $10 - 11i = (3 - 2i) * (4 - i)$ where each factor is prime.

Activity 3.21 Just above Activity 3.20 three different factorizations of $19 + 13i$ were given. Which one is the computer's choice?

Sometimes the last displayed factor is a unit. If we ask for a factorization of $10 - 12i$ the computer produces $10 - 12i = (1 + i)^2 * (6 + 5i) * (-1)$. Here the last factor, -1, is in white to emphasise that it is a unit and the factorization appears in this form purely as a consequence of the way in which the computer searches for factors. An expression in which every factor was prime could easily be obtained by multiplying the unit by one of the other prime factors. In this example we could have $10 - 12i = (1 + i)^2 * (-6 - 5i)$ or $10 - 12i = (1 + i) * (-1 - i) * (6 + 5i)$. Whenever the computer factorizes a composite Gaussian integer it helps us to locate the factors in the complex plane by changing their colours.

Activity 3.22 In Activity 3.16 we saw that the four numbers $-19 + i$, $-18 + i$, $-18 + 2i$ and $-19 + 2i$ are composite. Find complete factorizations of each one.

⟩ SUMMARY OF CHAPTER 3

After introducing algebraic numbers and algebraic integers this chapter concentrated on quadratic numbers. They are roots of quadratic polynomials in $\mathbb{Z}[x]$ and quadratic integers are roots of the monic quadratic polynomials in $\mathbb{Z}[x]$. Quadratic numbers are all of the form $a + b\sqrt{d} \in \mathbb{Q}(\sqrt{d})$ where a and b are rational and d is a square-free rational integer. When a and b are rational integers the number $a + b\sqrt{d}$ is a quadratic integer and if $d \not\equiv 1 \pmod 4$ there are no others in $\mathbb{Q}(\sqrt{d})$. However if $d \equiv 1 \pmod 4$ there is another class of quadratic integers comprising those numbers of the form $\frac{a+b\sqrt{d}}{2}$ where a and b are both odd. These two sorts of quadratic integer for $d \equiv 1 \pmod 4$ can be amalgamated as the one type of number $r + s\sigma$ where r and s are rational integers and $\sigma = \left(\frac{1+\sqrt{d}}{2}\right)$. In every case the integers of $\mathbb{Q}(\sqrt{d})$ form an integral domain.

Most of the chapter was taken up with exploring the Gaussian integers which are the quadratic integers in the field $\mathbb{Q}(\sqrt{-1})$. The computer enables us to move about among these quadratic integers and makes it easy to perform addition, subtraction, multiplication and division with them. As well as showing the results of arithmetic operations, the computer shows the value of the norm of each Gaussian integer. Geometrically the norm gives the square of the distance of the integer from the origin of the complex plane and it has the important property of being multiplicative. That is, the norm of a product is always equal to the product of the norms of the factors, as in (3.16). This multiplicative property of the norm made it easy to check, at the end of section 3.5, that the only invertible Gaussian integers are 1, -1, i and $-i$. Another very significant aspect of the norm which we met in section 3.5 is that we can always divide one Gaussian integer by another with a remainder that is either zero or has a smaller norm than the divisor. So the Gaussian integers form a Euclidean domain and we saw how the computer could help us to find GCDs using Euclid's algorithm.

The last section introduced irreducible and composite Gaussian integers. Irreducible elements play a central role since every non-zero and non-unit element is either irreducible or a product of irreducibles. There are infinitely many irreducible integers and we proved this using an analogue of Euclid's proof that there are infinitely many primes in $\mathbb{Z}$. It is a lot easier to see that there are also infinitely many composite Gaussian integers; and in that connection we deduced the interesting geometrical result that there are arbitrarily large squares in the complex plane which just contain composite integers. Irreducible integers also have the important prime property: if an irreducible Gaussian integer divides a product it has to divide one of the factors. The fact that the Gaussian irreducibles are all prime means that there is unique factorization in $\mathbb{Z}[i]$ and the computer will factorize those Gaussian integers that it displays.

〉 EXERCISES FOR CHAPTER 3

1. Find the minimal polynomials of $\sqrt{11} - \sqrt{7}$; $3 + \sqrt{-1}$; $\frac{6-2\sqrt{-7}}{5}$; $2 + 3\sqrt{7} - \sqrt{15}$.

2. Which of the following are algebraic integers $\frac{5+\sqrt{17}}{2}$; $\frac{5+\sqrt{17}}{3}$; $2\sqrt{5} - 13\frac{1}{3}$?

3. If a, b, $n \neq 0$ are rational integers find polynomials in $\mathbb{Z}[x]$ that have the following complex numbers as roots: $a + bi$; $a - bi$; $\frac{a-bi}{n}$; $\frac{a-bi}{n^{\frac{1}{3}}}$.

4. If $\alpha \neq 0$ is a root of the rational polynomial $a_n x^n + a_{n-1} x^{n-1} + \ldots + a_1 x + a_0$ write down a related polynomial in $\mathbb{Q}[x]$ that has $\frac{1}{\alpha}$ as a root.

5. If α is an algebraic number prove that $n\alpha$ will be an algebraic integer for some natural number n.

6. Verify that $\sqrt{2}$ is not an element of $\mathbb{Q}(\sqrt{3})$. (Ask yourself if there is any number in $\mathbb{Q}(\sqrt{3})$ whose square is 2.) Does $\mathbb{Q}(\sqrt{3})$ contain all of the roots of $x^3 - 1$?; of $4x^2 - 28x + 37$?

7. Find non-zero integers b, c such that $x^2 + bx + c$ is irreducible in $\mathbb{Q}[x]$ but composite in $\mathbb{Q}(\sqrt{5})[x]$.

8. Which of the numbers 13, 15, 20, 28, 32 can be norms of integers in $\mathbb{Z}[i]$?

9. If π is irreducible in $\mathbb{Z}[i]$ and is not an associate of $1 + i$ prove that the four units 1, -1, i, -1 lie in different congruence classes modulo π.

10. The Gaussian integers $-4 + 15i$ and $8 + 3i$ are at diagonally opposite corners of a square in the complex plane. What numbers are at the other two corners? If $-4 + 15i$ and $8 + 3i$ are at adjacent corners of a square what are the possibilities for the other two corners?

11. Suppose that $a + bi$ and $x + yi$ are in $\mathbb{Z}[i]$ with $x + yi \neq 0 + 0i$. Prove that we can always write

$$a + bi = (p_1 + q_1 i)(x + yi) + r_1 + s_1 i$$

with $p_1 + q_1 i$, $r_1 + s_1 i$ in $\mathbb{Z}[i]$ and $\frac{1}{2}\text{Norm}(x + yi) \leqslant \text{Norm}(r_1 + s_1 i) < \frac{3}{2}\text{Norm}(x + yi)$.

Chapter 4

〉 Arithmetic in quadratic domains

〉4.1 Multiplicative norms

As we have seen, the existence of a multiplicative norm in $\mathbb{Z}[i]$ is very important, for example in helping to decide which numbers are units and which are primes. It turns out that every domain of quadratic integers has a multiplicative norm, although we do not always have the additional bonus of the norm being Euclidean. First of all if $d \not\equiv 1 \pmod 4$, (3.9) shows that an integer α in $\mathbb{Q}(\sqrt{d})$ will be of the form $a + b\sqrt{d}$ for some integers a, b. If α' denotes $a - b\sqrt{d}$, the norm $N(\alpha)$ of α is then defined as

$$N(\alpha) = \alpha\alpha' = (a + b\sqrt{d})(a - b\sqrt{d}) = a^2 - db^2. \qquad (4.1)$$

When $d = -1$ this is exactly the same as our previous definition of the norm in $\mathbb{Z}[i]$. If $d \equiv 1 \pmod 4$, (3.8) shows that a typical integer α can be written as $\frac{a+b\sqrt{d}}{2}$ where a and b are either both odd or both even. Putting $\alpha' = \frac{a-b\sqrt{d}}{2}$, the norm of α is defined in this case by

$$N(\alpha) = \alpha\alpha' = \left(\frac{a + b\sqrt{d}}{2}\right)\left(\frac{a - b\sqrt{d}}{2}\right) = \frac{a^2 - db^2}{4}. \qquad (4.2)$$

As in Example 2.3, whenever d is not a square the elements of the field $\mathbb{Q}(\sqrt{d})$ are all of the form $r + s\sqrt{d}$ for rational numbers r and s. So we could extend the above definitions throughout $\mathbb{Q}(\sqrt{d})$ by defining the norm of $r + s\sqrt{d}$ to be the rational number

$$N(r + s\sqrt{d}) = (r + s\sqrt{d})(r - s\sqrt{d}) = r^2 - ds^2. \qquad (4.3)$$

It is then multiplicative because

$$N(r_1 + s_1\sqrt{d}) * N(r_2 + s_2\sqrt{d})$$
$$= (r_1 + s_1\sqrt{d})(r_1 - s_1\sqrt{d})(r_2 + s_2\sqrt{d})(r_2 - s_2\sqrt{d})$$
$$= (r_1 + s_1\sqrt{d})(r_2 + s_2\sqrt{d})(r_1 - s_1\sqrt{d})(r_2 - s_2\sqrt{d})$$
$$= [(r_1 r_2 + s_1 s_2 d) + (r_1 s_2 + r_2 s_1)\sqrt{d}]$$
$$* [(r_1 r_2 + s_1 s_2 d) - (r_1 s_2 + r_2 s_1)\sqrt{d}]$$
$$= N[(r_1 r_2 + s_1 s_2 d) + (r_1 s_2 + r_2 s_1)\sqrt{d}]$$
$$= N[(r_1 + s_1\sqrt{d})(r_2 + s_2\sqrt{d})] .$$

Just as in (3.16), this result can easily be extended to show that for any number of factors

the norm of any product in $\mathbb{Q}(\sqrt{d})$ *is equal to*

the product of the norms of the factors   (4.4)

Example 4.1
For any d equation (4.3) implies that the norm of the rational number $r = r + 0\sqrt{d}$ is just its square r^2. If d is negative the norm of $r + s\sqrt{d} = r + s\sqrt{|d|}i$ is $r^2 + |d|s^2$ which is the square of its distance from the origin of the complex plane.

Of particular interest to us is the fact that the norm of α is a rational integer whenever α is an integer in $\mathbb{Q}(\sqrt{d})$. Certainly in (4.1) $a^2 - db^2$ is an integer; and when $d \equiv 1 \pmod 4$ we have $a^2 - db^2 \equiv 1 - 1*1 \equiv 0 \pmod 4$, so that in (4.2) $\frac{a^2 - db^2}{4}$ is also an integer.

The computer will let us work in some other quadratic domains as well as in $\mathbb{Z}[i]$. For example, while in the Gaussian integer display, type '#−2' and press ENTER. The heading instantly changes to 'Prime Integers in $\mathbb{Q}(\sqrt{-2})$' and instead of the Gaussian integers the main display begins to show the integers of this new domain. As usual prime elements are shown in red and units in white and, as before, there is a yellow cross initially at the origin which we can move about the domain by using the cursor keys. The display here is again part of the complex plane where we move along the real axis in rational integer units and along the vertical, or imaginary, direction in multiples of $\sqrt{-2}$. Each quadratic integer here is of the form $a + b\sqrt{-2}$ for some rational integers a and b (see (3.9)), so this domain of integers can be described as the result of adjoining $\sqrt{-2}$ to $\mathbb{Z}$ and is written $\mathbb{Z}[\sqrt{-2}]$. As we move through the domain the side panel shows which integer the cross is currently centred on together with its norm.

Activity 4.1 What are the norms of the following integers of $\mathbb{Q}(\sqrt{-2})$: $3 + 7\sqrt{-2}$, $3 - 7\sqrt{-2}$, $\sqrt{-2}$, -1, $-1 + 2\sqrt{-2}$, $1 - 2\sqrt{-2}$?

In $\mathbb{Q}(\sqrt{-2})$ the norm of $a + b\sqrt{-2}$ is $a^2 + 2b^2$ and the product of $a + b\sqrt{-2}$ and $c + d\sqrt{-2}$ is $(ac - 2bd) + (ad + bc)\sqrt{-2}$. So the fact that the norm is multiplicative here is the same as the identity

$$(a^2 + 2b^2)(c^2 + 2d^2) = (ac - 2bd)^2 + 2(ad + bc)^2. \qquad (4.5)$$

In this domain we can use the computer to add, subtract, multiply and divide just as we did in $\mathbb{Z}[i]$. We move the cross-shaped cursor to the first of two chosen integers, press the key marked $+$, $-$, $*$ or $/$, depending on the operation we want, then move to the second integer and press ENTER. The result appears as before at the bottom of the display.

Activity 4.2 What are the products of the following pairs of quadratic integers?

(i) $7 + 2\sqrt{-2}$, $2 - \sqrt{-2}$; (ii) $-2 - 5\sqrt{-2}$, $-5 + 2\sqrt{-2}$; (iii) $1 + 2\sqrt{-2}$, $5 - 4\sqrt{-2}$; (iv) $-6 + 2\sqrt{-2}$, $7 - \sqrt{-2}$; (v) $3 + \sqrt{-2}$, $3 - \sqrt{-2}$.
In each case verify that the norm of the product is the product of the norms.

Activity 4.3 Show that in $\mathbb{Z}[\sqrt{-2}]$ the sum of $3 + 4\sqrt{-2}$ and $-10 + 7\sqrt{-2}$ is divisible by $1 + \sqrt{-2}$ but not by $1 - \sqrt{-2}$; and that the difference of $20 - 9\sqrt{-2}$ and $-14 + 8\sqrt{-2}$ is divisible by $-3 + 2\sqrt{-2}$ and also by $-3 - 2\sqrt{-2}$.

⟩4.2 Application of norms to units in quadratic domains

One of the main differences between the displays for $\mathbb{Z}[i]$ and for $\mathbb{Z}[\sqrt{-2}]$ is that 1 and -1 are the only units shown in $\mathbb{Z}[\sqrt{-2}]$ whereas there are four units 1, -1, i and $-i$ in $\mathbb{Z}[i]$. We originally used norms to help us decide which elements of $\mathbb{Z}[i]$ were invertible (at the end of section 3.5, just before Activity 3.11) and we can do the same not only in $\mathbb{Z}[\sqrt{-2}]$ but in any quadratic domain. Suppose first that υ is a unit in the domain of integers of $\mathbb{Q}(\sqrt{d})$ and that ψ is its inverse. Then

$$N(\upsilon)N(\psi) = N(\upsilon\psi) = N(1) = 1.$$

That means that the rational integer $N(\upsilon)$ is a divisor of 1, so it has to be ± 1. On the other hand suppose that α is a quadratic integer in $\mathbb{Q}(\sqrt{d})$ with norm ± 1. We use α' to denote the algebraic conjugate of α as in (4.1) and (4.2), and recall that $\alpha\alpha' = N(\alpha)$. Then

$$\alpha * \alpha' N(\alpha) = (N(\alpha))^2 = 1,$$

so that $\alpha' N(\alpha)$, which is an integer in $\mathbb{Q}(\sqrt{d})$, is the inverse of α. Thus α is a unit. So

an algebraic integer α is a unit in $\mathbb{Q}(\sqrt{d})$

if and only if the norm of α is ± 1.   (4.6)

In order to see why this implies that there are different numbers of units in some cases we need to go a little further and bring in the formulae (4.1) and (4.2). If $d \not\equiv 1 \pmod 4$ it is clear from (4.1) and (4.6) that the typical integer $a + b\sqrt{d}$ in $\mathbb{Q}(\sqrt{d})$ will be a unit precisely when

$$a^2 - db^2 = \pm 1.$$ (4.7)

If $d \equiv 1 \pmod 4$ the integer $\frac{a+b\sqrt{d}}{2}$ will be a unit whenever $\frac{a^2-db^2}{4} = \pm 1$, which is the same as

$$a^2 - db^2 = \pm 4.$$ (4.8)

These equations always have some rational integer solutions. For instance the equation (4.7) has the solutions $a = 1, b = 0$ and $a = -1, b = 0$. They correspond to the units 1 and -1 and for some values of d, such as $d = -2$, there are no other solutions. For each d we shall find the exact number of solutions of (4.7) and (4.8) and thus the number of units in the domain of integers in $\mathbb{Q}(\sqrt{d})$. We begin with the negative values of d as in those cases we can say what all the solutions are. Note that when d is negative the expression $a^2 - db^2 = a^2 + |d|b^2$ can never be negative. Consequently norms must be zero or positive (see (4.1) and (4.2)); and also we can only hope to solve equations (4.7) and (4.8) when their right hand sides are $+1$ or $+4$ respectively.

When d is negative and not congruent to 1 modulo 4 equation (4.7) amounts to $a^2 + |d|b^2 = \pm 1$; so if $|d| \geqslant 2$ it can only be satisfied with $b = 0$ and $a = \pm 1$ (and $+1$ on the right side of the equation). If $d = -1$, the equation is $a^2 + b^2 = \pm 1$, and the solutions are $a = \pm 1, b = 0$ and $a = 0, b = \pm 1$. So in $\mathbb{Z}[i]$ (the case $d = -1$) there are four units ± 1,

$\pm i$ as we already know, and for each other negative $d \not\equiv 1 \pmod 4$ there are only the two units ± 1.

When d is negative and congruent to 1 modulo 4, equation (4.8) is $a^2 + |d|b^2 = \pm 4$. So if $|d| > 4$ the only solutions are $a = \pm 2$ and $b = 0$ giving the units $\frac{2+0\sqrt{d}}{2} = 1$ and $\frac{-2+0\sqrt{d}}{2} = -1$. The only negative $d \equiv 1 \pmod 4$ with $|d| \leqslant 4$ is $d = -3$ and then there are six solutions: $a = \pm 2$, $b = 0$; $a = \pm 1$, $b = \pm 1$. These give six corresponding units in the domain of quadratic integers in $\mathbb{Q}(\sqrt{-3})$, namely $\frac{2+0\sqrt{-3}}{2} = 1$, $\frac{-2+0\sqrt{-3}}{2} = -1$, $\frac{1+\sqrt{-3}}{2}$, $\frac{1-\sqrt{-3}}{2}$, $\frac{-1+\sqrt{-3}}{2}$ and $\frac{-1-\sqrt{-3}}{2}$.

So the situation as regards units in $\mathbb{Z}[\sqrt{-2}]$ is not at all exceptional, since when d is negative the domain of quadratic integers in $\mathbb{Q}(\sqrt{d})$ nearly always has just two units. The exceptions are $\mathbb{Z}[i]$ with four units and the integers of $\mathbb{Q}(\sqrt{-3})$ where there are six units. In $\mathbb{Q}(\sqrt{-3})$ we know from (3.11) that every integer is a linear combination, with rational integer coefficients, of 1 and $\sigma = \left(\frac{1+\sqrt{-3}}{2}\right)$. In other words any quadratic integer there can be expressed as $a + b\sigma$, for some rational integers a and b, and so can be reached by moving a steps along the real axis and then b steps in the σ-direction. We say that 1 and σ form an *integral basis* for the quadratic integers in $\mathbb{Q}(\sqrt{-3})$.

Activity 4.4 Move into $\mathbb{Q}(\sqrt{-3})$ by typing '#-3' (and pressing ENTER of course). The six units, $\sigma = \left(\frac{1+\sqrt{-3}}{2}\right)$, $-1 + \sigma$, -1, $-\sigma$, $1 - \sigma$, 1, are indicated in white and we can perform arthmetic here in exactly the same way as in $\mathbb{Z}[i]$ and $\mathbb{Z}[\sqrt{-2}]$. For instance, to multiply $1 + 8\sigma$ by σ we move the yellow cross to $1 + 8\sigma$, press '$*$', and then move to σ and press ENTER. Find all the associates (the unit multiples) of $1 + 8\sigma$ by multiplying it in turn by each of the six units.

Activity 4.5 Multiply σ by itself to obtain σ^2. Continue to multiply by σ and thus show that each unit is one of the powers σ, σ^2, σ^3, σ^4, σ^5, $\sigma^6 = 1$.

Activity 4.6 Find which unit is the inverse of σ. Do this either by noticing that $\sigma * \sigma^5 = \sigma^6 = 1$, or by using the computer to find $1/\sigma$. Which unit is the inverse of $\sigma - 1$?

When d is positive norms can be either positive or negative and we can have units that have norm -1 as well as those with norm $+1$. You can experience this among the integers of $\mathbb{Q}(\sqrt{2})$ by typing '#2'. Here each integer is of the form $a + b\sqrt{2}$ with norm $a^2 - 2b^2$. So the norm will be negative whenever b^2 is greater than $a^2/2$. For instance, the norm of $11 + 9\sqrt{2}$ is $11^2 - 2 * 9^2 = -41$ and the norm of $7 + 5\sqrt{2}$ is $7^2 - 2 * 5^2 = -1$ so that, from (4.6), $7 + 5\sqrt{2}$ is a unit. Note that every element of $\mathbb{Q}(\sqrt{2})$, and in particular every quadratic integer in $\mathbb{Q}(\sqrt{2})$, is real. So in this case the display is not a representation of the complex plane; but the computer writes multiples of $\sqrt{2}$ vertically to aid us in recognizing the two parts of each quadratic integer.

Activity 4.7 Divide 1 by $7 + 5\sqrt{2}$ and so find its inverse.

Activity 4.8 How many units are visible on your monitor? Which of them have norm -1?

In order to find all the units for $d > 0$ we again have to investigate the equations (4.7) and (4.8). We prove in Appendix 5 that if D is any positive rational integer that is not a square then there are infinitely many solutions of both $a^2 - Db^2 = +1$ and $a^2 - Db^2 = +4$. In particular if $d > 1$ is a square-free integer then each of the equations $a^2 - db^2 = +1$ and $a^2 - db^2 = +4$ has infinitely many solutions. So in $\mathbb{Q}(\sqrt{d})$ there are infinitely many units with norm $+1$. There may be no solutions of $a^2 - db^2 = -1$ or $a^2 - db^2 = -4$, and so no units with norm -1, but when there are any there must be infinitely many. In every case there will be a unique unit $\eta > 1$ such that every unit can be written as $\pm\eta^m$ for some integer m which may be positive, negative or zero. The unit η is called the *fundamental unit* and it will have norm 1 if every unit has norm $+1$. If there are units with norm -1 then η has norm -1 and all the units of norm -1 are of the form $\pm\eta^{2k-1}$, while the units with norm $+1$ are of the form $\pm\eta^{2k}$. When $d \not\equiv 1 \pmod 4$, η can be written as $t_1 + u_1\sqrt{d}$ and (t_1, u_1) is then the fundamental solution of $t^2 - du^2 = 1$ if η has norm 1, or of $t^2 - du^2 = -1$ if η has norm -1. Similarly, when $d \equiv 1 \pmod 4$, η can be written as $\frac{1}{2}(t_1 + u_1\sqrt{d})$ and then (t_1, u_1) is the fundamental solution of $t^2 - du^2 = 4$ if η has norm 1, or of $t^2 - du^2 = -4$ if η has norm -1. (See Appendix 5; especially equations (A5.27), (A5.30), (A5.36) and (A5.38).)

Activity 4.9 The unit $\eta = 1 + \sqrt{2}$ is the fundamental unit in $\mathbb{Q}(\sqrt{2})$ because $(1, 1)$ is the fundamental solution of $a^2 - 2b^2 = -1$ (see Appendix 5). In $\mathbb{Q}(\sqrt{2})$ select a unit, other than ± 1, that is visible on your monitor and use the computer to show that it is either a power of η or minus a power of η.

Activity 4.10 Find the fundamental unit in the domain of integers of

$\mathbb{Q}(\sqrt{3})$. [Integers here are of the form $a + b\sqrt{3}$ with norms of the form $a^2 - 3b^2$. Use congruences modulo 3 to show that there are no units with norm -1 in this domain. So the fundamental unit corresponds to the fundamental solution of $a^2 - 3b^2 = 1$.]

Activity 4.11 What is the fundamental unit in the domain of integers of $\mathbb{Q}(\sqrt{5})$?

⟩**4.3 Irreducible and prime quadratic integers**

In this section we shall see that there are infinitely many irreducible integers in any quadratic domain. The first step towards this result is to note that every non-zero integer, which is not a unit, must be divisible by some irreducible integer. To see this suppose that α is a non-zero integer in a quadratic domain $\mathcal{D}$ and that it is not a unit. Consider all the non-unit divisors of α in $\mathcal{D}$. They include α itself, and their norms can only range over the finite number of rational integer divisors of $|\text{Norm}(\alpha)|$ (and not necessarily all of those). Let β be a non-unit divisor of α, the absolute value of whose norm is as small as possible. Then β must be irreducible. Otherwise it would be a product of two quadratic integers, γ and δ say, neither of which was a unit. So $\beta = \gamma\delta$ would mean that γ was a divisor of α, and then $1 < |\text{Norm}(\gamma)| < |\text{Norm}(\beta)|$ would contradict our choice of β as a divisor of α with smallest possible norm.

Example 4.2
In $\mathbb{Q}(\sqrt{3})$ the integer $-19 + 17\sqrt{3}$ has norm $-506 = -2 * 11 * 23$. It has non-unit divisors of norms -2, -11, -23, 22, 46, 253 and -506. For example $1 + \sqrt{3}$, $1 + 2\sqrt{3}$, $2 - 3\sqrt{3}$, $7 + 3\sqrt{3}$, $-7 - \sqrt{3}$, $-16 + \sqrt{3}$, $-19 + 17\sqrt{3}$ or any of their infinitely many associates. The previous paragraph assures us that each divisor of norm -2 is irreducible, and indeed the computer emphasises $1 + \sqrt{3}$ in red.

Activity 4.12 Use congruences modulo 3 to show that no integer in $\mathbb{Q}(\sqrt{3})$ could have norm $+2$, $+11$, $+23$, -22, -46, -253 or $+506$.

Activity 4.13 In the domain of integers in $\mathbb{Q}(\sqrt{-5})$ each element is of the form $a + b\sqrt{-5}$, with norm $a^2 + 5b^2$. The integer $17 + 5\sqrt{-5}$ has norm $414 = 2 * 3^2 * 23$ so each of its divisors has to have a positive norm that divides 414. Use the computer to verify that $1 + \sqrt{-5}$ is a divisor of norm 6. Then use congruences modulo 5 to show that 2 and 3 are not allowable norms here. Thus $1 + \sqrt{-5}$ has smallest norm of all the divisors of $17 + 5\sqrt{-5}$ and is therefore an irreducible divisor.

Activity 4.14 Use the result of the first paragraph of this section to show that, in any quadratic domain, if the absolute value of a quadratic integer's norm is a rational prime then that integer is irreducible. (See also Example 3.5 and following remarks.)

We could now show that there are always infinitely many irreducible integers in the domain of integers in $\mathbb{Q}(\sqrt{d})$ (whether d is positive or negative) by adapting Euclid's proof, just as we did in the domain of Gaussian integers. Another interesting way is to try to construct an infinite sequence of non-zero quadratic integers, none of which is a unit, and every two of which are relatively prime. Suppose in other words, that we could find a sequence of non-zero and non-unit integers $\alpha_0, \alpha_1, \alpha_2, \ldots$ with the property that no two members of the sequence have divisors in common (apart from unit divisors). Then each term α_i would either be irreducible itself or be divisible by some other irreducible; but no irreducible integer dividing α_i could also divide any other term α_j. So, if we then took an irreducible element dividing α_0, say ϖ_0, an irreducible element dividing α_1, say ϖ_1, and so on, the sequence $\varpi_0, \varpi_1, \varpi_2, \ldots$ would be an infinite sequence of distinct irreducible integers. Perhaps surprisingly it is easy to write down some suitable sequences. One such is the sequence of Fermat numbers. This is the sequence of rational integers $F_0, F_1, \ldots$ which we can define recursively by the relations

$$F_0 = 3 \quad \text{and} \quad F_n - 2 = F_{n-1}(F_{n-1} - 2) \text{ for } n \geqslant 1. \qquad (4.9)$$

It is not hard to see (using induction for example) that these are all rational integers so they occur in whatever quadratic domain we are working in. Also every member of the sequence is odd and is at least 3; so none of them is zero and none is a unit for they each have norm greater than 1. It is actually easy to give an explicit formula for each term, as a simple induction proof using (4.9) shows that $F_n = 2^{2^n} + 1$ for every n. However, of more importance to us at the moment, is the next formula that we can also use induction to deduce from (4.9)

$$F_n - 2 = F_{n-1}F_{n-2}\ldots F_0. \qquad (4.10)$$

Suppose now that two terms F_i and F_j are both divisible by a quadratic integer δ, where the subscripts i, j are such that $i > j$. Then, from (4.10),

$$F_i - 2 = F_{i-1}\ldots F_j \ldots F_0$$

so that δ must divide $2 = F_i - (F_{i-1} \ldots F_j \ldots F_0)$. But F_i, being odd, must be of the form $2K + 1$ for some rational integer K. That is, $1 = F_i - 2K$, and both F_i and 2 are divisible by δ. So δ must also divide 1 and is therefore a unit. Thus each pair of Fermat numbers is relatively prime, in the sense that they have no irreducible divisors in common. So in any quadratic domain we can make up an unending sequence of irreducible integers ϖ_0, ϖ_1, ϖ_2, ... by taking ϖ_0 to be any irreducible divisor of F_0, ϖ_1 to be an irreducible divisor of F_1, ... and ϖ_i to be an irreducible divisor of F_i for $i > 1$. Since every two Fermat numbers are relatively prime in our quadratic domain, no two elements of the sequence ϖ_0, ϖ_1, ... can be equal or even be associates.

Individual Fermat numbers are difficult to investigate because they grow very rapidly. The first few are $F_0 = 3$, $F_1 = 5$, $F_2 = 17$, $F_3 = 257$ and $F_4 = 65537$. However the computer will help us to examine F_0, F_1, F_2 in various domains of integers. For instance in $\mathbb{Q}(\sqrt{2})$, F_0 and F_1 are irreducible and $F_2 = (5 + 2\sqrt{2})(5 - 2\sqrt{2})$. So we know from our general analysis that, in $\mathbb{Q}(\sqrt{2})$, 3, 5 and $5 + 2\sqrt{2}$ are different irreducible integers which cannot be associates.

Activity 4.15 In the domain of Gaussian integers, find three irreducible integers that divide 3, 5, 17 respectively.

In any quadratic domain the pattern of irreducible integers possesses some symmetry. If a quadratic integer α is irreducible then so is its algebraic conjugate α' (see (4.1), (4.2)) for $\alpha = \beta\gamma$ is equivalent to $\alpha' = \beta'\gamma'$. Also if ν is a unit then $\nu\alpha$ has the same irreducible or composite character as α. In particular $-\alpha$ and α are either both irreducible or both composite. When the computer is first calculating the irreducible integers in a new domain and finds an irreducible integer α it fills in $-\alpha$, α' and $-\alpha'$ at the same time as α. In $\mathbb{Z}[i]$ it fills in up to 8 irreducible values each time because it multiplies each irreducible and its conjugate by i and $-i$ as well as by -1 and 1; and in the domain of integers in $\mathbb{Q}(\sqrt{-3})$ it fills in up to 12 values at once as it multiplies each newly found irreducible by the six units as well as taking conjugates.

We know from chapter 3 that every irreducible Gaussian integer has the prime property: if an irreducible element divides a product then it must divide one of the factors. As we shall see, this is true of irreducible integers in several other domains. In $\mathbb{Z}[i]$ it follows from the fact that the domain is Euclidean and in the next section we prove that there are other values of d for which the integers of $\mathbb{Q}(\sqrt{d})$ form a Euclidean domain. There are also domains of quadratic integers which are not Euclidean domains but where every irreducible integer nevertheless has the prime property. Examples are the domains of integers in $\mathbb{Q}(\sqrt{-19})$, $\mathbb{Q}(\sqrt{-43})$, $\mathbb{Q}(\sqrt{-67})$ and $\mathbb{Q}(\sqrt{-163})$, although we shall not prove that (see e.g. Stewart and Tall (1979)).

Section 3 of Appendix 1 shows that being irreducible is a necessary pre-condition for being prime but there are many domains containing irreducible integers that do not have the prime property. An example is the domain of integers in $\mathbb{Q}(\sqrt{-5})$. In Activity 4.13 we saw that $1+\sqrt{-5}$ is irreducible in that domain and divides $17 + 5\sqrt{-5}$. Thus $1 + \sqrt{-5}$ divides the product $(1 + 3\sqrt{-5}) * (2 - \sqrt{-5}) = 17 + 5\sqrt{-5}$. But it does not divide either of the factors $1 + 3\sqrt{-5}$ or $2 - \sqrt{-5}$; for the norm of $1 + \sqrt{-5}$, namely 6, does not divide either Norm$(1 + 3\sqrt{-5}) = 46$ or Norm$(2 - \sqrt{-5}) = 9$.

⟩4.4 Euclidean domains of quadratic integers

We saw in chapter 3 that the integers of $\mathbb{Q}(\sqrt{-1})$ form a Euclidean domain, with the ordinary distance norm acting as Euclidean function. It turns out that the integers of $\mathbb{Q}(\sqrt{d})$ form a Euclidean domain for several other values of d, with the Euclidean function being the absolute value of the multiplicative norm defined in (4.1) and (4.2). If α is an integer in $\mathbb{Q}(\sqrt{d})$, and $N(\alpha)$ is its multiplicative norm, then certainly

(i) $|N(\alpha)| \geqslant 0$ for any non-zero α. (Indeed $|N(\alpha)| > 0$ for a non-zero α as, with d not a square, $a^2 - db^2$ cannot be zero unless both a and b are zero.)

(ii) If $\beta | \alpha$ then $|N(\beta)| \leqslant |N(\alpha)|$, because if $\alpha = \beta\gamma$ then $|N(\alpha)| = |N(\beta)| * |N(\gamma)|$.

So, as in section 1.4 of Appendix 1, in order to show that $|N(\ )|$ is a Euclidean function we have to establish that for integers α and $\beta \neq 0$ in $\mathbb{Q}(\sqrt{d})$ we have integers κ and γ in $\mathbb{Q}(\sqrt{d})$ such that

(iii) $\alpha = \kappa\beta + \gamma$ with either $\gamma = 0$ or $|N(\gamma)| < |N(\beta)|$.

We shall establish this for several small values of d by modelling our argument on the one we used for $\mathbb{Z}[\sqrt{-1}]$ in section 3.5. Suppose first that $d \not\equiv 1 \pmod 4$ and that $\alpha = a + b\sqrt{d}$, $\beta = x + y\sqrt{d}$ are typical integers in $\mathbb{Q}(\sqrt{d})$ with $\beta \neq 0$. In $\mathbb{Q}(\sqrt{d})$ we have

$$\frac{a + b\sqrt{d}}{x + y\sqrt{d}} = \frac{(a + b\sqrt{d})(x - y\sqrt{d})}{(x + y\sqrt{d})(x - y\sqrt{d})} = \frac{(a + b\sqrt{d})(x - y\sqrt{d})}{x^2 - dy^2}$$

so we try to divide the numerator $(a + b\sqrt{d})(x - y\sqrt{d}) = (ax - dby) + (bx - ay)\sqrt{d}$ by the norm $N = x^2 - dy^2$ of $x + y\sqrt{d}$. Using ordinary division with remainder in $\mathbb{Z}$ we can certainly write $ax - dby = N*p + R$ for some rational integers p and R with $|R| \leqslant \frac{|N|}{2}$; and likewise $bx - ay = N*q + S$ for integers q and S with $|S| \leqslant \frac{|N|}{2}$. Then

$$(a + b\sqrt{d})(x - y\sqrt{d}) = N * (p + q\sqrt{d}) + R + S\sqrt{d} \qquad (4.11)$$

or equivalently

$$\frac{a + b\sqrt{d}}{x + y\sqrt{d}} = p + q\sqrt{d} + \frac{R + S\sqrt{d}}{N}. \qquad (4.12)$$

Since $N = (x - y\sqrt{d})(x + y\sqrt{d})$, (4.11) shows that $R + S\sqrt{d} = (a + b\sqrt{d})(x - y\sqrt{d}) - N * (p + q\sqrt{d})$ is exactly divisible by $x - y\sqrt{d}$. If $R + S\sqrt{d} = (x - y\sqrt{d})(r + s\sqrt{d})$ for integers r and s, then from either (4.11) or (4.12)

$$(a + b\sqrt{d}) = (p + q\sqrt{d}) * (x + y\sqrt{d}) + r + s\sqrt{d} \qquad (4.13)$$

and

$$\left|\text{Norm}(r + s\sqrt{d})\right| = \left|\text{Norm}(R + S\sqrt{d})\right| / \left|\text{Norm}(x - y\sqrt{d})\right| = \frac{|R^2 - dS^2|}{|N|}.$$

Using the inequalities satisfied by R, S we see that for $d = -2, -1, 2$ and 3 this gives

$$\left|\text{Norm}(r + s\sqrt{d})\right| \leqslant \frac{3N^2/4}{|N|} = \frac{3|N|}{4} = \frac{3}{4}\left|\text{Norm}(x + y\sqrt{d})\right|. \qquad (4.14)$$

When $d \equiv 1 \pmod 4$ the argument is similar, but now the typical integers α and β are of the form $\frac{a + b\sqrt{d}}{2}$ and $\frac{x + y\sqrt{d}}{2}$ respectively. So $\frac{\alpha}{\beta}$ is

$$\frac{a + b\sqrt{d}}{x + y\sqrt{d}} = \frac{(a + b\sqrt{d})(x - y\sqrt{d})}{(x + y\sqrt{d})(x - y\sqrt{d})} = \frac{[(a + b\sqrt{d})(x - y\sqrt{d})]/4}{(x^2 - dy^2)/4}$$

where $N = (x^2 - dy^2)/4$ is the integer norm of $\frac{x + y\sqrt{d}}{2}$. The numerator here is $\left(\frac{a + b\sqrt{d}}{2}\right)\left(\frac{x - y\sqrt{d}}{2}\right) = \left(\frac{ax - dby}{2}\right) + \left(\frac{bx - ay}{2}\right)\sqrt{d}$ which is an integer of $\mathbb{Q}(\sqrt{d})$; so $\frac{ax-dby}{2}$ and $\frac{bx-ay}{2}$ are rational integers and we can certainly choose rational integers q, S to satisfy $\frac{bx-ay}{2} = N * q + S$ with $|S| \leqslant \frac{|N|}{2}$ whence

$$\frac{bx-ay}{4} = N * \left(\frac{q}{2}\right) + \frac{S}{2} \quad \text{with} \quad \left|\frac{S}{2}\right| \leqslant \frac{|N|}{4}. \tag{4.15}$$

We can also choose rational integers p' and R' such that $\frac{ax-dby}{2} = N * p' + R'$ where $|R'| \leqslant \frac{|N|}{2}$. We can even ensure that p' and q are congruent modulo 2, at the expense of allowing R' to lie in a larger range. If they are not congruent modulo 2 and if $0 \leqslant R' \leqslant \frac{|N|}{2}$ then we subtract N from R' and add 1 to p', getting $\frac{ax-dby}{2} = N*(p'+1)+(R'-N)$ with $-N \leqslant R'-N \leqslant -\frac{|N|}{2}$. If $p' \not\equiv q \pmod 2$ and $-\frac{|N|}{2} \leqslant R' < 0$ we add N to R' and subtract 1 from p', obtaining $\frac{ax-dby}{2} = N*(p'-1)+(R'+N)$ with $\frac{|N|}{2} \leqslant R'+N < N$. Supposing that these adjustments have been made if necessary, we then have integers p, R $(= p', R'; p'+1, R'-N;$ or $p'-1, R'+N)$ such that $p \equiv q \pmod 2$ and

$$\frac{ax-dby}{2} = N * p + R \quad \text{where} \quad |R| \leqslant |N|$$

or

$$\frac{ax-dby}{4} = N * \left(\frac{p}{2}\right) + \frac{R}{2} \quad \text{with} \quad \left|\frac{R}{2}\right| \leqslant \left|\frac{N}{2}\right|. \tag{4.16}$$

Hence

$$\left(\frac{a+b\sqrt{d}}{2}\right)\left(\frac{x-y\sqrt{d}}{2}\right) = N * \left(\frac{p+q\sqrt{d}}{2}\right) + \frac{R+S\sqrt{d}}{2} \tag{4.17}$$

where $\frac{p+q\sqrt{d}}{2}$ is an integer in $\mathbb{Q}(\sqrt{d})$. Then $\frac{R+S\sqrt{d}}{2} = \left(\frac{a+b\sqrt{d}}{2}\right)\left(\frac{x-y\sqrt{d}}{2}\right) - N*\left(\frac{p+q\sqrt{d}}{2}\right)$ is also a quadratic integer in $\mathbb{Q}(\sqrt{d})$ and is exactly divisible by $\frac{x-y\sqrt{d}}{2}$. If $\frac{R+S\sqrt{d}}{2} = \left(\frac{x-y\sqrt{d}}{2}\right)\left(\frac{r+s\sqrt{d}}{2}\right)$ for an integer $\frac{r+s\sqrt{d}}{2}$, then from (4.17)

$$\frac{a+b\sqrt{d}}{2} = \left(\frac{p+q\sqrt{d}}{2}\right) * \left(\frac{x+y\sqrt{d}}{2}\right) + \frac{r+s\sqrt{d}}{2} \tag{4.18}$$

and from the inequalities satisified by R, S we see that

$$\left|\operatorname{Norm}\left(\frac{r+s\sqrt{d}}{2}\right)\right| = \left|\operatorname{Norm}\left(\frac{R+S\sqrt{d}}{2}\right)\right| \Big/ \left|\operatorname{Norm}\left(\frac{x-y\sqrt{d}}{2}\right)\right|$$
$$= \frac{|(\frac{R}{2})^2 - d(\frac{S}{2})^2|}{|N|} \leqslant \frac{N^2/4 + |d|N^2/16}{|N|}.$$

For $d = -11, -7, -3, 5$ this implies that

$$\left|\operatorname{Norm}\left(\frac{r+s\sqrt{d}}{2}\right)\right| < \left|\operatorname{Norm}\left(\frac{x+y\sqrt{d}}{2}\right)\right|. \qquad (4.19)$$

The equations (4.13) and (4.18), together with the inequalities in (4.14) and (4.19), express the 'division with remainder of smaller norm' property that we need. The domain of integers in $\mathbb{Q}(\sqrt{d})$ is therefore a Euclidean domain, with the Euclidean norm being the absolute value of the multiplicative norm, for each of $d = -11, -7, -3, -2, -1, 2, 3$ and 5. It can be proved that there are no other quadratic Euclidean domains with negative d, and that when d is positive there are 13 other Euclidean domains with the analogous Euclidean function (see e.g. Stewart and Tall (1979) and also Stark (1979)).

The computer will help us to explore the quadratic Euclidean domains with $-3 \leqslant d \leqslant 5$ and whenever it has to perform division it will always choose a remainder of smaller norm than the divisor. Note that when $d \not\equiv 1 \pmod 4$ it writes the result (4.13) of a typical division with remainder as

$$(a+b\sqrt{d})/(x+y\sqrt{d}) = p + q\sqrt{d} + \frac{(r+s\sqrt{d})}{(x+y\sqrt{d})}.$$

When $d \equiv 1 \pmod 4$ it uses a similar display, but writes each quadratic integer $\frac{a+b\sqrt{d}}{2}$ in the form $A + B\sigma$ where $A = \frac{a-b}{2}$, $B = b$ and $\sigma = \left(\frac{1+\sqrt{d}}{2}\right)$. It is always easy to recover the alternative form since the side panel tells us the corresponding values of a and b for the current quadratic integer.

Activity 4.16 In $\mathbb{Q}(\sqrt{-3})$ use the computer in each of the following cases to divide the first number by the second. Check that the norm of each remainder has a smaller absolute value than that of the corresponding divisor. Here σ denotes $\left(\frac{1+\sqrt{-3}}{2}\right)$.

(i) $1 + 12\sigma$, $12 - 6\sigma$; (ii) $29 - 12\sigma$, $1 - 17\sigma$; (iii) $17 - 13\sigma$, $\sigma - 15$; (iv) $-14 + 8\sigma$, $-7 + 2\sigma$; (v)$21 - 22\sigma$, $15 + 10\sigma$.

As we saw earlier in the case of $\mathbb{Z}[i]$ there may be more than one possibility for a remainder with smaller norm. For instance if we stay in $\mathbb{Q}(\sqrt{-3})$ and try to divide 19σ by 7 we could either have $19\sigma = 7 * 3\sigma + (-2\sigma)$ with Norm$(-2\sigma) = 4 <$Norm$(7) = 49$, or $19\sigma = 7 * 2\sigma + (5\sigma)$ with Norm$(5\sigma) = 25 <$Norm(7).

Activity 4.17 Repeat Activity 4.16 for the same numbers defined in terms of σ, but with the domain being the integers of $\mathbb{Q}(\sqrt{5})$ and σ denoting $\left(\frac{1+\sqrt{5}}{2}\right)$. Depending on your graphics driver, you may have to alter the scale of the picture on the screen in order to move the screen cursor to some numbers such as $29 - 12\sigma$. You can do that by holding down the key marked 'CTRL' and, while doing so, pressing the 'S' key. Repeating the same key presses will restore the original scale.

If we know that division is always possible with a remainder whose norm has smaller absolute value than that of the divisor, we can be assured that when Euclid's algorithm is applied to two non-zero integers it will always terminate since the absolute values of the norms of the remainders are decreasing. Then, as we saw in sections 1.4 and 3.5, the last non-zero remainder in Euclid's algorithm will be a greatest common divisor in the sense of being a common divisor that is divisible by every other common divisor. So any two quadratic integers in one of these Euclidean domains will have a GCD which can be found using Euclid's idea of systematic division with remainder.

Activity 4.18 In $\mathbb{Q}(\sqrt{-3})$ use Euclid's algorithm to find greatest common divisors of each of the pairs of integers in Activity 4.16. Switch to $\mathbb{Q}(\sqrt{5})$ and find greatest common divisors of the same pairs defined in terms of $\sigma = \left(\frac{1+\sqrt{5}}{2}\right)$. Are the answers necessarily the same?

In any quadratic domain where there are greatest common divisors each pair of integers will have more than one GCD, because if γ is a GCD of α and β and ν is a unit then $\nu\gamma$ will also be a GCD (see Appendix 1). For negative values of d there will be at most 6 greatest common divisors of a given pair of integers (and only that many when $d = -3$), but when d is positive each pair of quadratic integers will have an infinite number of different greatest common divisors.

Activity 4.19 In $\mathbb{Q}(\sqrt{3})$ use Euclid's algorithm to find greatest common divisors of the following pairs of quadratic integers. (i) $9+11\sqrt{3}, 3-6\sqrt{3}$; (ii) $-17+12\sqrt{3}, 2-10\sqrt{3}$; (iii) $5-27\sqrt{3}, 23$; (iv) $31+2\sqrt{3}, -3+17\sqrt{3}$. In each case find at least eight different greatest common divisors by multiplying your first GCD by different units.

In any Euclidean domain every irreducible element is prime (see Appendix 1, section 5). We also saw at the end of the previous section that in the domain of integers in $\mathbb{Q}(\sqrt{-5})$ there are irreducible integers that are not prime. So that domain cannot be Euclidean. In it there are still going to be common divisors for each pair of non-zero integers since units are always common divisors; but the next example shows that there might not always be a greatest common divisor.

Example 4.3
Use the computer in $\mathbb{Q}(\sqrt{-5})$ to show that $1 + 5\sqrt{-5}$ and $-8 + 2\sqrt{-5}$ are each divisible by $3 + \sqrt{-5}$ and by $4 - \sqrt{-5}$ in the domain of integers here. If there were another common divisor which was divisible by each of these common divisors then its norm would have to be divisible by both $\text{Norm}(3 + \sqrt{-5}) = 14$ and by $\text{Norm}(4 - \sqrt{-5}) = 21$. Thus the norm of any greatest common divisor would have to be divisible by 42 and so, to divide into $-8 + 2\sqrt{-5}$ of norm 84, it would have to have norm 42 or 84. However no integer of norm 84 can divide into $1+5\sqrt{-5}$ since that has norm 126; and congruences modulo 5 show that there are no integers of norm 42 in $\mathbb{Z}[\sqrt{-5}]$. So $1 + 5\sqrt{-5}$ and $-8+2\sqrt{-5}$ have no common divisor that is divisible by every common divisor.

⟩4.5 Factorization into irreducible integers in quadratic domains

In any integral domain we classify each non-zero and non-unit element as either irreducible or composite; and in all the quadratic domains each composite integer is a product of a finite number of irreducible integers. This is true no matter whether the quadratic domain is Euclidean or not and the general proof is almost identical to that given in section 3.6 for the Gaussian integers. The main difference is that we must allow for norms perhaps being negative.

We use induction on the absolute values of the norms in a domain $\mathcal{D}$ of quadratic integers, and we know that all the integers in $\mathcal{D}$ with norm ± 1 must be units. Suppose we have proved that for some rational integer $n \geqslant 1$ every non-zero element in $\mathcal{D}$ whose norm is between $-n$ and

n is either a unit, irreducible or a product of irreducibles (as we have for $n = 1$). If there are no further composite elements we have proved what we want. Otherwise let k be the least rational integer greater than n which can occur as the absolute value of the norm of a composite element, and let α be a composite quadratic integer in $\mathcal{D}$ whose norm is either k or $-k$. (So k may be $n + 1$ but may be larger if $n + 1, \ldots, k - 1$, and $-n - 1, \ldots, -k + 1$, only occur—if at all—as norms of irreducible elements.) Then, by definition, α will be the product of two integers β and γ in $\mathcal{D}$, neither of which is a unit, and $|N(\beta)| * |N(\gamma)| = k$. Thus $1 < |N(\beta)| < k$ and $1 < |N(\gamma)| < k$, which means that each of β, γ is either irreducible or, if composite, the absolute value of its norm is at most n. In any case, the inductive assumption implies that β and γ are each either irreducible or can be written as a product of irreducibles. Writing $\beta = \beta_1 \ldots \beta_r$ and $\gamma = \gamma_1 \ldots \gamma_s$, where each β_i and γ_j is irreducible, gives $\alpha = \beta_1 \ldots \beta_r \gamma_1 \ldots \gamma_s$. So we can always make the inductive step to elements of larger norm and the result, that each composite integer is a product of a finite number of irreducible integers, follows in all cases.

We know that in a Euclidean domain every irreducible element has the prime property—that if an irreducible element π divides a product $\alpha\beta$ it must divide either α or β. The general treatment of Euclidean domains in Appendix 1 shows that this implies that in those domains every non-zero element is either a unit or can be expressed essentially uniquely as a product of irreducible elements. (Actually an almost identical proof to that at the end of section A1.5, but using the absolute value of the multiplicative norm, shows that uniqueness of factorization holds in any quadratic domain where irreducibles are primes, irrespective of whether the domain is Euclidean or not.) For instance, in $\mathbb{Q}(\sqrt{-3})$ with $\sigma = \left(\frac{1+\sqrt{-3}}{2}\right)$, we could factorize the integer $8 + 11\sigma$ as

$$(1 + \sigma)(2 + \sigma)(4 - \sigma)$$

or

$$(-1 + 2\sigma)(4\sigma - 3)(-2 - \sigma)$$

or as

$$(1 + 3\sigma)(-1 + 3\sigma)(-1 + 2\sigma)$$

where each term is irreducible and the last two factorizations have been obtained from the first by multiplying each factor by a unit and rearranging the terms. These are the only sorts of ways in which we can alter the factorization. In this case, if the first factor $1 + \sigma$ is multiplied by a unit ν_1 and the next by a unit ν_2, then the last factor $4 - \sigma$ has to be multiplied by $(\nu_1\nu_2)^{-1}$ to keep the product the same. As there are 6 different choices for ν_1 and 6 for ν_2, there will be 36 choices for the pair of units ν_1, ν_2. Also the terms of each factorization can be rearranged in 6 different ways so altogether there are $36 * 6 = 216$ ways of writing down what amounts to the same factorization each time. Similarly in any quadratic domain with $d < 0$ there will only be a finite number of seemingly different ways of writing down a given factorization. However any quadratic domain with $d > 0$ has an infinite number of units so that, even without rearranging the terms, there will be an infinite number of ways of altering a given factorization to make it look different. For example in $\mathbb{Z}[\sqrt{2}]$ the number $8 + 11\sqrt{2}$ could be written as $(4 + 3\sqrt{2})(17 - 10\sqrt{2})$ or $(10 - 7\sqrt{2})(117 + 83\sqrt{2})$ or $(24 + 17\sqrt{2})(91 - 64\sqrt{2})$ just by multiplying $4 + 3\sqrt{2}$ by a different unit each time (and $17 - 10\sqrt{2}$ by the same unit's inverse).

Activity 4.20 Use the computer to verify that $10 - 7\sqrt{2}$ and $24 + 17\sqrt{2}$ can be obtained by multiplying $4 + 3\sqrt{2}$ by suitable units while $117 + 83\sqrt{2}$ and $91 - 64\sqrt{2}$ can be obtained by dividing $17 - 10\sqrt{2}$ by the same units.

The computer will automatically find a factorization of any composite quadratic integer if you want it to. Assuming that you are working in $\mathbb{Z}[\sqrt{2}]$, you could factorize the number $8 + 11\sqrt{2}$ above by simply moving the on-screen cursor to that number and pressing ENTER. In that instance the display at the bottom of the screen would show
$$8 + 11\sqrt{2} = \sqrt{2} * (3 - 7\sqrt{2}) * (-1 - \sqrt{2})$$
where the first two factors would be shown in red to indicate that they are prime and the last factor $-1 - \sqrt{2}$ would be in white to show that it is a unit. A factorization in which every factor was prime could then be derived by combining the unit with one of the other terms. For example, since $(3 - 7\sqrt{2}) * (-1 - \sqrt{2}) = 11 + 4\sqrt{2}$, we have $8 + 11\sqrt{2} = \sqrt{2} * (11 + 4\sqrt{2})$ where each factor is prime. This is equivalent to the first factorization in the previous paragraph, because $(4 + 3\sqrt{2}) = \sqrt{2} * (3 + 2\sqrt{2})$ and $(17 - 10\sqrt{2}) = (11 + 4\sqrt{2})/(3 + 2\sqrt{2})$. The computer sometimes has to put a unit at the end of a factorization because it prefers choosing prime factors with coefficients that are small. If you accidentally asked the computer to factorize a quadratic integer shown in red, such as $11 - 5\sqrt{2}$, it would say '$11 - 5\sqrt{2}$ is prime'. Similarly if you asked it to factorize an integer shown in white, such as $-7 + 5\sqrt{2}$, it would put '$-7 + 5\sqrt{2}$' is a unit. It will ignore requests to

factorize zero!

Activity 4.21 Factorize the following numbers into prime factors in $\mathbb{Z}[\sqrt{2}]$: (i) $24+8\sqrt{2}$; (ii) $22+14\sqrt{2}$; (iii) 32; (iv) $31-9\sqrt{2}$; (v) $-12+5\sqrt{2}$.

Activity 4.22 Factorize the following numbers into prime factors in $\mathbb{Z}[\sqrt{3}]$: (i) $-28-4\sqrt{3}$; (ii) $2-19\sqrt{3}$; (iii) $-34+10\sqrt{3}$; (iv) $-43-12\sqrt{3}$; (v) $-38-20\sqrt{3}$; (vi) $31+2\sqrt{3}$.

In most cases it will be obvious if any of the quadratic integers have prime factors in common. However if you are testing two integers for common factors by factorizing each one instead of using Euclid's algorithm, it is as well to be careful if you cannot immediately see any common factors. In $\mathbb{Z}[\sqrt{2}]$ for instance, the integers $60+24\sqrt{2}$ and $37+8\sqrt{2}$ have the prime factorizations

$$60+24\sqrt{2} = 3 * (\sqrt{2})^4 * (5+2\sqrt{2})$$

and

$$37+8\sqrt{2} = (9+7\sqrt{2}) * (13-11\sqrt{2})$$

but they still have the common factor $5+2\sqrt{2}$. This is because $9+7\sqrt{2}$ is an associate of $5+2\sqrt{2}$, being $(5+2\sqrt{2})*(1+\sqrt{2})$. Here we would have been alerted to this possibility by looking at the norms of the prime factors. The three different primes occurring as factors of $60+24\sqrt{2}$ have norms 9, -2 and 17 respectively, while the two given factors of $37+8\sqrt{2}$ have norms -17 and -73. So the only two which might be unit multiples of each other are $5+2\sqrt{2}$ and $9+7\sqrt{2}$, and we can test that by dividing one into the other. The two numbers $60+24\sqrt{2}$ and $37+8\sqrt{2}$ have no other prime factors in common, so $5+2\sqrt{2}$ is a greatest common divisor. In practice we would be unlikely to be mislead if we asked the computer to factorize $60+24\sqrt{2}$ and $37+8\sqrt{2}$ because the computer will put a given prime factor in the same form every time rather than picking different associates on different occasions.

Activity 4.23 What are the computer's factorizations of $37+8\sqrt{2}$ and $60+24\sqrt{2}$ in $\mathbb{Z}[\sqrt{2}]$?

Activity 4.24 In $\mathbb{Z}[\sqrt{3}]$ factorize each of the quadratic integers occurring in Activity 4.19. Use these factorizations to give greatest common factors of each of the pairs of integers.

We have seen near the beginning of this section that when every irreducible quadratic integer is prime there is unique factorization of composite integers into irreducible factors (apart from multiplication by units and the rearrangment of terms). The converse is also true. That is, if essentially unique factorization of composites into irreducibles holds throughout a certain quadratic domain, then in that domain every irreducible integer is prime. In order to see this, suppose that, in some quadratic domain, every composite integer can be factored in essentially only one way into a product of irreducible integers and that ϖ is irreducible. Then to show that ϖ is prime we must show that if ϖ divides a product $\alpha\beta$ it has to divide either α or β. If α is a unit then ϖ divides an associate of β and so divides β. Similarly if β is a unit ϖ will divide α. Otherwise both α and β can be written as products of irreducible elements, say $\alpha = \rho_1\rho_2\ldots\rho_r$ and $\beta = \sigma_1\sigma_2\ldots\sigma_s$ where each ρ_i and σ_j is ireducible. So $\alpha\beta = \rho_1\rho_2\ldots\rho_r\sigma_1\sigma_2\ldots\sigma_s$ and that must be the unique expression for $\alpha\beta$ as a product of irreducible elements. But $\alpha\beta$ can also be written as $\varpi * (\frac{\alpha\beta}{\varpi})$ and here $\frac{\alpha\beta}{\varpi}$ will be a product of irreducibles, say $\frac{\alpha\beta}{\varpi} = \tau_1\tau_2\ldots\tau_t$. Then we will have

$$\alpha\beta = \rho_1\rho_2\ldots\rho_r\sigma_1\sigma_2\ldots\sigma_s = \varpi\tau_1\tau_2\ldots\tau_t.$$

Because of the uniqueness of factorization, each irreducible element in the product $\varpi\tau_1\tau_2\ldots\tau_t$ must be an associate of one of the terms in the product $\rho_1\ldots\rho_r\sigma_1\ldots\sigma_s$. In particular, ϖ will either divide one of ρ_1, $\rho_2, \ldots, \rho_r$ or one of $\sigma_1, \sigma_2, \ldots, \sigma_s$. In the first case it divides α and in the second it divides β.

In $\mathbb{Z}[\sqrt{-5}]$ we have already seen an instance of an irreducible integer that is not prime. So in that domain we should now expect there to be at least one composite integer that has two genuinely different factorizations into irreducible integers. In fact there are many, and one is the number $17 + 5\sqrt{-5}$ which we met in Activity 4.13 and at the end of section 4.3.

Activity 4.25 Verify that

$$17 + 5\sqrt{-5} = (1 + \sqrt{-5})(7 - 2\sqrt{-5}) = (1 + 3\sqrt{-5})(2 - \sqrt{-5}). \quad (4.20)$$

The factors occurring in these two products have norms 6, 69, 46 and 9 respectively. Use congruences modulo 5 to show that norms in $\mathbb{Z}[\sqrt{-5}]$ (which are all of the form $a^2 + 5b^2$) can never be congruent to 2 or 3 modulo 5. That implies that 2, 3, 23 cannot be norms and so none of those four factors could be expressed as a product of two non-unit integers. In other words they are each irreducible. Also the fact that their norms are different means that no two are associates. So the products in (4.20) are completely different factorizations of $17+5\sqrt{-5}$ into irreducible integers.

The computer can automatically factor any composite integer in the part of $\mathbb{Z}[\sqrt{-5}]$ that is displayed on the screen and, if we want, it will also search for an alternative factorization. Suppose for instance that we wanted the computer to factorize $7 - 7\sqrt{-5}$. As usual we should move the on-screen cross to $7 - 7\sqrt{-5}$ and press the ENTER key. The computer would then show

$7 - 7\sqrt{-5} = (3 + \sqrt{-5}) * (1 + 2\sqrt{-5}) * (-1)$

If we now pressed ENTER again (without moving the cursor), the factorization would change to

$7 - 7\sqrt{-5} = (1 - \sqrt{-5}) * 7$

We could of course absorb the unit -1 into either of the first pair of factors to give for example $7 - 7\sqrt{-5} = (-3 - \sqrt{-5}) * (1 + 2\sqrt{-5})$. These four factors, $1 - \sqrt{-5}$, 7, $-3 - \sqrt{-5}$ and $1 + 2\sqrt{-5}$, have norms 6, 49, 14 and 21 and, using the information about the residue classes of possible norms in Activity 4.25, it is easy to see that none of them can be divisible by any non-unit integer of smaller norm. So they are all irreducible and no two of them can be associates.

Some numbers only have one factorization and, in those cases, pressing ENTER twice will just make the computer repeat the same factors. Sometimes the computer repeats the same factors but gives them in a different order, so you should take a note of the factors each time before concluding that you have found two different factorizations.

Activity 4.26 For which of the following numbers does the computer produce different factorizations?
(i) $19 - 4\sqrt{-5}$; (ii) $-12 - 4\sqrt{-5}$; (iii) $4 - 14\sqrt{-5}$; (iv) $12 - 10\sqrt{-5}$; (v) $-30 + 13\sqrt{-5}$.
In each case where the computer gives different sets of factors, use your knowledge of norms to verify that the factorizations do not just differ by unit multiples.

Activity 4.27 How many rational integers can you find that can each be split into irreducible factors in more than one way?

There are also numbers that have more than two different factorizations. For example, the number $63 - 21\sqrt{-5}$ can be written as

$$3 * 7 * (3 - \sqrt{-5})$$

or as

$$(-3 - \sqrt{-5}) * (4 + \sqrt{-5}) * (1 + 2\sqrt{-5})$$

or as

$$(1 + \sqrt{-5}) * (4 - \sqrt{-5}) * (2 - 3\sqrt{-5}).$$

However, in these cases the computer will only give two of the possible factorizations.

Activity 4.28 Show that each of the nine factors of $63 - 21\sqrt{-5}$ listed above is irreducible. Show further that none of them is a unit multiple of any other. (Remember that 1 and -1 are the only units in $\mathbb{Z}[\sqrt{-5}]$.)

⟩ SUMMARY OF CHAPTER 4

Here we first met the general definition of the norm in a quadratic domain and saw that the norm is always multiplicative. The computer will display integers in various quadratic domains together with their norms and will help to do arithmetic with the quadratic integers. One of the benefits of the norm being multiplicative is that it simplifies deciding which integers are units. Thus an algebraic integer is a unit if and only if its norm is ± 1. By investigating this condition in detail for each d we found how many units there were among the integers of $\mathbb{Q}(\sqrt{d})$. There are only two units for negative d apart from $d = -1$ and $d = -3$ when there are 4 and 6 units respectively. Whereas if d is positive there are infinitely many units.

Irreducible integers are defined as in any integral domain (see Appendix 1.3) and we showed that in every domain of quadratic integers there are infinitely many irreducibles. They do not always possess the prime property, but will certainly do so whenever the domain is Euclidean; and in section 4.4 we proved that the integers of $\mathbb{Q}(\sqrt{d})$ form a Euclidean domain, with the absolute value of the usual multiplicative norm as Euclidean norm, for $d = -11, -7, -3, -2, -1, 2, 3$ and 5. One consequence is that in these cases we can use Euclid's algorithm to find a greatest common divisor of two integers. Another is that in these domains every irreducible integer is prime, which means there is unique factorization of composite integers into irreducible factors. In section 4.5 we also saw the converse implication. Namely, that if in any domain there is always unique factorization into irreducible elements then in that domain every irreducible element must be prime. The computer will automatically factorize many composite quadratic integers in $\mathbb{Q}(\sqrt{d})$ for $|d| \leqslant 5$. In $\mathbb{Z}[\sqrt{-5}]$, where different products of irreducibles can be the same, it will try to find alternative factorizations.

⟩ EXERCISES FOR CHAPTER 4

1. Which of the numbers 7, 8, 12, 15, 28, 41 can be norms of integers in $\mathbb{Q}(\sqrt{-2})$? Which do you think can be norms of integers in $\mathbb{Q}(\sqrt{2})$?

2. What are the fundamental units in $\mathbb{Q}(\sqrt{6})$, $\mathbb{Q}(\sqrt{7})$?

3. Adapt Euclid's proof to prove that there are infinitely many irreducible integers in $\mathbb{Q}(\sqrt{-3})$. (See the proof for $\mathbb{Z}[i]$ in section 3.6.)

4. In the domain of integers of $\mathbb{Q}(\sqrt{-5})$ show that both 3 and $1+\sqrt{-5}$ are common divisors of 12 and $-9+3\sqrt{-5}$. Prove though that these last two integers have no greatest common divisors.

5. Prove that for any $n \geqslant 1$ the Fermat number $2^{2^n}+1$ is composite in $\mathbb{Z}[i]$. Is this true in $\mathbb{Z}[\sqrt{3}]$?

6. In the domain of integers of $\mathbb{Q}(\sqrt{6})$ the number 6 factorizes as $2*3$ and also as $\sqrt{6}*\sqrt{6}$. Why does this not contradict unique factorization in $\mathbb{Q}(\sqrt{6})$?

7. Prove that the integers of $\mathbb{Q}(\sqrt{-6})$ do not possess unique factorization. [Hint: look at -6 itself.] Prove more generally that if p and q are different primes the integers of $\mathbb{Q}(\sqrt{-pq})$ do not possess unique factorization.

8. The set $\mathbb{Z}[\sqrt{-3}]$ consists of all the numbers of the form $a+b\sqrt{-3}$ where a and b are rational integers. Prove that this set together with the usual addition and multiplication of complex numbers is an integral domain (all you need to show is that it is closed under these operations and contains 0 and 1). So it is a sub-domain of the domain of all quadratic integers in $\mathbb{Q}(\sqrt{-3})$. Units and irreducible elements are defined in $\mathbb{Z}[\sqrt{-3}]$ as in any integral domain (see Appendix 1, sections 2,3). How many units are there here? Prove that 4 has two distinct factorizations into irreducible elements in $\mathbb{Z}[\sqrt{-3}]$. Are your factorizations distinct in the domain of all quadratic integers in $\mathbb{Q}(\sqrt{-3})$?.

9. Let D be the set of polynomials with rational coefficients but with
    no linear term in x. So each polynomial in D is of the form
    $a + bx^2 + cx^3 + \ldots$. Prove that D is an integral domain. Prove
    further that the elements x^2 and x^3 are irreducible in D. Use them
    to give two distinct factorizations of x^6 in D. (This example of
    non-unique factorization has been given in a more general context
    in Chapman (1992) and in Anderson and Pruis (1991).)

10. Generalize the domain in the previous example by considering the
    set of polynomials with rational coefficients but no terms in x, x^2,
    $\ldots, x^{n-1}$ for a given $n > 1$. Prove that this set is an integral domain.
    By choosing n appropriately show that for any N there are integral
    domains in which there are composite elements having more than N
    different expressions as products of irreducibles.

11. Suppose that for some d every irreducible integer in $\mathbb{Q}(\sqrt{d})$ is
    known to be prime, but it is not known whether the integers of
    $\mathbb{Q}(\sqrt{d})$ form a Euclidean domain. Prove nevertheless that there
    must be unique factorization of composite integers into products of
    irreducibles in this domain. [Follow the proof in Appendix 1, section
    5 but use the absolute value of the multiplicative norm in $\mathbb{Q}(\sqrt{d})$
    without assuming it is a Euclidean norm.]

〉 Chapter 5

〉 Composite rational integers and sums of squares

〉5.1 Rational primes

Every quadratic integer divides its own norm because the norm of the quadratic integer α is the product of α and its algebraic conjugate (see (4.1) and (4.2)). Also each norm is a rational integer, so if we just look at the quadratic factors of rational integers we eventually meet every quadratic integer. Note that if α divides the rational integers m and n it will divide any linear combination of them. So it must divide their greatest common divisor g, as there will be integers r, s in $\mathbb{Z}$ with $g = rm + sn$. In particular, if two rational integers have greatest common divisor 1 in $\mathbb{Z}$ then, in any quadratic domain, they can only have units as common divisors.

Activity 5.1 In $\mathbb{Z}[\sqrt{3}]$ verify that 10 and 14 are each divisible by $19 - 11\sqrt{3}$. So their GCD 2 is also divisible by $19 - 11\sqrt{3}$. (Incidentally, $19 - 11\sqrt{3}$ does not appear as a factor if you ask the computer to factorize 2 in this domain. Check that $19 - 11\sqrt{3}$ is related to one of the factors that does appear.)

We can go further if unique factorization always holds in the quadratic domain, for then each irreducible quadratic integer is also prime. So if π is a prime integer with norm $n = p_1^{e_1} \ldots p_r^{e_r}$, it must divide one of the rational prime factors $p_1, \ldots, p_r$ of its norm. In this case every irreducible quadratic integer occurs as a factor of a rational prime.

Example 5.1
Any two different rational primes will have greatest common divisor 1 in $\mathbb{Z}$. As in the first paragraph, they can only have unit divisors in common

108

in any quadratic domain. So we cannot have an irreducible quadratic integer that is a factor of more than one rational prime.

Any rational integer that is composite in $\mathbb{Z}$ certainly remains composite in any quadratic domain; for a factorization in $\mathbb{Z}$ is also a factorization— not necessarily into prime factors—in a quadratic domain. On the other hand, in each of the domains we have looked at there are some rational primes that are irreducible integers in the domain but also some that are composite. Our first aim will be to try to decide which rational primes remain irreducible (or equivalently, which are composite) in any given quadratic domain. Suppose then that p is a rational prime that is composite in the domain of integers of $\mathbb{Q}(\sqrt{d})$. So p will be divisible by an integer α with $1 < |\text{Norm}(\alpha)| < \text{Norm}(p) = p^2$ and, since $\text{Norm}(\alpha)$ divides $\text{Norm}(p)$, this must mean that the norm of α is either p or $-p$. On the other hand if β is a quadratic integer whose norm is either p or $-p$ (and whose conjugate is denoted by β'), we will either have $\text{Norm}(\beta) = \beta\beta' = p$, or $\text{Norm}(\beta) = \beta\beta' = -p$ making $\beta(-\beta') = p$. Thus

> *p is composite if and only if there is a quadratic*
>
> *integer whose norm is either p or $-p$.*   (5.1)

If $d \not\equiv 1 \pmod 4$ norms are of the form $a^2 - db^2$, so p will be composite if and only if there are rational integers a, b such that

$$a^2 - db^2 = \pm p \qquad (5.2)$$

and, if $d \equiv 1 \pmod 4$, p will be composite if and only if there are rational integers a, b such that $a \equiv b \pmod 2$ and

$$\frac{a^2 - db^2}{4} = \pm p. \qquad (5.3)$$

These equations can frequently be shown to have no solutions by using simple congruence arguments. For example, when $d = -1$, equation (5.2) is

$$a^2 + b^2 = p \qquad (5.4)$$

since $-p$ could not be a norm. Now, modulo 4, squares can only be congruent to 0 or 1 and therefore $a^2 + b^2$ can never be 3 modulo 4. So if $p \equiv 3 \pmod 4$ there cannot be any Gaussian integer with norm p and so p will be irreducible in $\mathbb{Z}[i]$. The rational primes 3, 7, 11, 19, ... are thus irreducible Gaussian integers. Again, when $d = -3$, equation (5.3) becomes

$$a^2 + 3b^2 = 4p \qquad (5.5)$$

which implies $a^2 \equiv p$ (mod 3). But every square is congruent to 0 or 1 modulo 3, so if $p \equiv 2$ (mod 3) then (5.5) can never be satisfied and p will be irreducible in $\mathbb{Q}(\sqrt{-3})$. This applies for instance to 2, 5, 11, 17, ...

Activity 5.2 When $d = -5$, equation (5.2) is $a^2 + 5b^2 = p$. Use congruences modulo 5 to show that this cannot be satisfied with integers a, b if $p \equiv 2, 3$ (mod 5). Thus 2, 3, 7, 13, 17, 23, ... are all irreducible in $\mathbb{Z}[\sqrt{-5}]$. When $d = 5$, equation (5.3) is $a^2 - 5b^2 = \pm 4p$. Use congruences modulo 5 again to show that, whichever sign is taken, this equation has no solutions if $p \equiv 2, 3$ (mod 5). Thus 2, 3, 7, 13, 17, 23, ... are also irreducible in the domain of integers in $\mathbb{Q}(\sqrt{5})$. [We shall see later that this describes all the positive rational primes that are irreducible in $\mathbb{Q}(\sqrt{5})$. However, there are other rational primes that are irreducible in the domain of integers in $\mathbb{Q}(\sqrt{-5})$ (see Example 5.2 below).]

Another useful congruence approach, at least when p is odd, is to consider (5.2) and (5.3) modulo p. Firstly if p were an odd prime and divided b in either of these cases, it would also have to divide a, and the left-hand side would then be divisible by p^2 and so could not be equal to $\pm p$. So, if p is odd, both (5.2) and (5.3) give

$$a^2 \equiv db^2 \ (\text{mod } p) \ \text{ and } \ b \not\equiv 0 \ (\text{mod } p). \qquad (5.6)$$

Any rational integer b appearing in a solution of (5.6) will thus have an inverse, say c, modulo p. So (5.6) implies

$$(ac)^2 \equiv db^2c^2 \equiv d \ (\text{mod } p). \qquad (5.7)$$

This means that if p is to be composite in the domain of integers in $\mathbb{Q}(\sqrt{d})$ then d has to be a square modulo p. Equivalently,

> *if d is not a square modulo p then the rational*
>
> *prime p will be irreducible in $\mathbb{Q}(\sqrt{d})$.* $\qquad (5.8)$

If p is 2 this is a vacuous statement since every d is a square modulo 2 (if d is odd then $d \equiv 1^2$ (mod 2) and if d is even then $d \equiv 0^2$ (mod 2)).

Example 5.2
The number -5 is not a square modulo 11. (This can be verified by simply squaring the integers from 0 to 5 inclusive as their squares represent all the square residue classes modulo 11.) So 11 will be an irreducible integer in $\mathbb{Q}(\sqrt{-5})$. Similarly it is straightforward to check that 3 is not a square modulo 17, so 17 will be irreducible in the domain of integers of $\mathbb{Q}(\sqrt{3})$.

We can actually go further than (5.8) because

if d is not a square modulo p then p will remain prime in $\mathbb{Q}(\sqrt{d})$. (5.9)

To see this, suppose that d is not a square modulo p, which necessarily means that p cannot be 2, and suppose first that $d \not\equiv 1 \pmod 4$ so that the integers in $\mathbb{Q}(\sqrt{d})$ are all of the form $a + b\sqrt{d}$. Say p divides a product $(a + b\sqrt{d})(x + y\sqrt{d}) = ax + dby + (ay + bx)\sqrt{d}$, but does not divide the factor $a + b\sqrt{d}$. Then the coefficients $ax + dby$ and $ay + bx$ must each be multiples of p, so p will divide $x(ax + dby) - dy(ay + bx) = a(x^2 - dy^2)$, and hence either a or $x^2 - dy^2$. If p divided a then, since it divides $ax + dby$ and $ay + bx$, it would have to divide dby and bx; but it can not divide b, as otherwise it would divide $a + b\sqrt{d}$, and it does not divide d as d is not a square modulo p. So in that case p would divide both x and y and thus $x + y\sqrt{d}$. The other possibility is that p divides $x^2 - dy^2$. Then if $y \not\equiv 0 \pmod p$ it would have an inverse y_1 modulo p, giving $(xy_1)^2 \equiv d \pmod p$ and contradicting d being a non-square modulo p. So, in this case y, and so also x, would have to be divisible by p and again $p | x + y\sqrt{d}$. The reasoning that leads to (5.9) is entirely similar if $d \equiv 1 \pmod 4$. If p then divided a product $(\frac{a+b\sqrt{d}}{2})(\frac{x+y\sqrt{d}}{2}) = \frac{\frac{ax+dby}{2} + (\frac{ay+bx}{2})\sqrt{d}}{2}$ it would have to divide each of the rational integers $\frac{ax+dby}{2}$ and $\frac{ay+bx}{2}$. This would again mean that p would divide $2x(\frac{ax+dby}{2}) - 2dy(\frac{ay+bx}{2}) = a(x^2 - dy^2)$. Then $p \nmid \frac{a+b\sqrt{d}}{2}$ and d not a square modulo p would lead as before to $p | \frac{x+y\sqrt{d}}{2}$.

The converse of (5.9) always holds if p is odd. For suppose that d is a square modulo an odd rational prime p. In other words $t^2 \equiv d \pmod p$ for some integer t. So in $\mathbb{Z}$ we have $t^2 - d = pq$ for a rational integer q. But this last equation is also an equation in the domain of integers of $\mathbb{Q}(\sqrt{d})$; so in that domain p divides $t^2 - d = (t - \sqrt{d})(t + \sqrt{d})$. However p does not divide either of the factors $t - \sqrt{d}$ or $t + \sqrt{d}$, as it does not divide the coefficients, $+1$ or -1, of $\sqrt{d}$. This shows that

> *if d is a square modulo an odd then*
>
> *p does not have the prime property in* $\mathbb{Q}(\sqrt{d})$.   (5.10)

The converse of the statement (5.8) would be that *if d is a square modulo p then p is composite in the domain of integers in* $\mathbb{Q}(\sqrt{d})$. This is not always true, even for odd primes, as we can see by taking $d = -5$ and $p = 23$. Here -5 is a square modulo 23 since $15^2 \equiv -5$ (mod 23); but 23 is irreducible in $\mathbb{Q}(\sqrt{-5})$ (see Activity 5.2). However in a quadratic domain with unique factorization, having the prime property is equivalent to being irreducible (see the end of section 3 in Appendix 1 and the paragraph just after Activity 4.24). So in such a domain (5.10) means that if d is a square modulo p then p is composite. Combining (5.8) and (5.10) we therefore get:

> *if there is unique factorization in* $\mathbb{Q}(\sqrt{d})$ *then*
>
> *an odd p is composite if and only if d is a square mod p.* (5.11)

When there is uniqueness of factorization this equivalence reduces the problem of finding whether p is composite in a quadratic domain to a problem entirely within the domain of rational integers. In any particular case we could, at worst, test whether d is a square modulo p by squaring the integers between 0 and $\frac{p-1}{2}$ inclusive and seeing whether one of them is congruent to d. This is very impractical when p is at all large but luckily there is a much quicker way to find if d is a square modulo p.

⟩5.2 Quadratic residues and the Legendre symbol

In this section we shall deal only with ordinary rational integers. First of all an integer a is called a *quadratic residue* of a prime p if it is congruent to a non-zero square modulo p. So a is a quadratic residue when we can find an integer x with

$$a \equiv x^2 (\text{mod } p) \quad \text{and} \quad x \not\equiv 0 \ (\text{mod } p). \tag{5.12}$$

If a is not divisible by p and is not a square modulo p it is a *quadratic non-residue* modulo p. Any integer that is divisible by p is certainly a square modulo p since it will be congruent to 0^2, but the term quadratic residue is reserved for the non-zero squares. Modulo 2, all the odd numbers are quadratic residues since they are each congruent to 1^2 (mod

2). They all belong to the same residue class so we say that there is only one quadratic residue modulo 2. When p is odd the numbers $1, 2, \ldots,$ $p - 1$ represent all possible non-zero residue classes so their squares certainly include all possible square residues. Among these squares, $(p-1)^2, \ldots, \left(\frac{p+1}{2}\right)^2$ are congruent to $1^2, \ldots, \left(\frac{p-1}{2}\right)^2$ respectively, since $x^2 \equiv (-x)^2$ for each x. Also no two of the squares $1^2, 2^2, \ldots, \left(\frac{p-1}{2}\right)^2$ are congruent because $x^2 \equiv y^2$ (mod p) implies $x^2 - y^2 = (x-y)(x+y) \equiv 0$ (mod p) and so either $y \equiv x$ or $y \equiv -x$ modulo p. So there are exactly $\frac{p-1}{2}$ different quadratic residues, represented by the squares $1^2, 2^2, \ldots,$ $\left(\frac{p-1}{2}\right)^2$. The remaining non-zero residues are the quadratic non-residues so there must be $\frac{p-1}{2}$ of them as well.

If p is an odd prime and a is an integer, the *Legendre symbol* $\left(\frac{a}{p}\right)$ is defined by

$$\left(\frac{a}{p}\right) = \begin{cases} +1 & \text{if } a \text{ is a quadratic residue modulo } p \\ -1 & \text{if } a \text{ is a quadratic non-residue modulo } p \\ 0 & \text{if } a \text{ is divisible by } p. \end{cases} \quad (5.13)$$

The advantage of defining this symbol is that it can be manipulated more easily than the original concept which often has to be described in words. For instance we always have

$$\left(\frac{a}{p}\right) = \left(\frac{b}{p}\right) \quad \text{if } a \equiv b \text{ (mod } p\text{)} \quad (5.14)$$

because the solubility, or insolubility, of (5.12) is unaffected if a is replaced by a congruent number b.

Example 5.3
For any odd prime p

$$\sum_{a=1}^{p-1} \left(\frac{a}{p}\right) = 0. \quad (5.15)$$

This is a translation into symbols of the fact that between 1 and $p - 1$ there are equal numbers of quadratic residues and non-residues. So the above sum contains equal numbers of terms that are $+1$ and terms that are -1, giving a total of zero.

Example 5.4
If k is not divisible by the odd prime p then

$$\left(\frac{k^2}{p}\right) = +1. \tag{5.16}$$

This is because, with $a = k^2$, (5.12) is satisfied with $x = k$.

We shall see that the Legendre symbol satisfies several other properties that enable its value to be calculated easily in every case.

If a and b are both quadratic residues then so is their product, because if $a \equiv x^2$ and $b \equiv y^2$ then $ab \equiv (xy)^2$ (mod p). This is the same as the statement that if $\left(\frac{a}{p}\right) = \left(\frac{b}{p}\right) = 1$ then $\left(\frac{ab}{p}\right) = 1$. To see what happens when we multiply a quadratic residue by a quadratic non-residue, suppose that a is a quadratic residue and consider the products $a, 2a, \ldots,$ $(p-1)a$ formed by multiplying a by each non-zero residue in turn. These are all distinct modulo p since $ax \equiv ay$ (mod p) implies $x \equiv y$ (mod p). We also know that each multiplication of a with a quadratic residue gives a quadratic residue; and since there are $\frac{p-1}{2}$ such products, they must give all the quadratic residues. The remaining products, of a by the quadratic non-residues, must therefore be congruent in some order to the $\frac{p-1}{2}$ quadratic non-residues. In other words if $\left(\frac{a}{p}\right) = +1$ and $\left(\frac{b}{p}\right) = -1$ then $\left(\frac{ab}{p}\right) = -1 = \left(\frac{a}{p}\right)\left(\frac{b}{p}\right)$. Interchanging the names of the elements a and b shows that $\left(\frac{ab}{p}\right)$ is also -1 if $\left(\frac{a}{p}\right) = -1$ and $\left(\frac{b}{p}\right) = +1$. Finally, consider the set of products $b, 2b, \ldots, (p-1)b$ where b is a quadratic non-residue modulo p. We have just seen that a product of a quadratic non-residue with a quadratic residue is a non-residue, so all the non-residues are represented by the products of b with all of the quadratic residues. So the other $\frac{p-1}{2}$ products, of b with each of the non-residues, must be all the quadratic residues. Thus if $\left(\frac{a}{p}\right) = -1$ and $\left(\frac{b}{p}\right) = -1$ then $\left(\frac{ab}{p}\right) = +1$. These cases show that when neither a nor b is divisible by p we have

$$\left(\frac{ab}{p}\right) = \left(\frac{a}{p}\right)\left(\frac{b}{p}\right). \tag{5.17}$$

This remains true if either a or b is divisible by p since then both sides are zero.

Example 5.5
We can sometimes using properties (5.14), (5.16) and (5.17) together to evaluate particular Legendre symbols. To find the value of $\left(\frac{33}{83}\right)$ for instance, we can argue that

$$\left(\frac{33}{83}\right) = \left(\frac{-50}{83}\right) \text{ using (5.14)}$$

$$= \left(\frac{-2}{83}\right)\left(\frac{25}{83}\right) \text{ using (5.17)}$$

$$= \left(\frac{-2}{83}\right) \text{ from (5.16)}$$

$$= \left(\frac{81}{83}\right) \text{ using (5.14) again}$$

$$= +1 \text{ from (5.16).}$$

This means that 33 must be a square modulo 83, and it is much easier to see that by manipulating Legendre symbols as above than by trying to find a number whose square is congruent to 33. In fact $38^2 \equiv 33$ (mod 83).

Any Legendre symbol can actually be calculated directly using a theorem of Fermat. If $p \nmid r$ Fermat's result (A1.15) states that $r^{p-1} \equiv 1$ (mod p). This implies

$$(r^{\frac{p-1}{2}} - 1)(r^{\frac{p-1}{2}} + 1) \equiv 0 \pmod{p}$$

and so

$$r^{\frac{p-1}{2}} \equiv \pm 1 \pmod{p}. \tag{5.18}$$

If r is a quadratic residue we must have the plus sign in (5.18) because if $r \equiv s^2$ (mod p) then

$$r^{\frac{p-1}{2}} \equiv s^{p-1} \equiv 1 = \left(\frac{r}{p}\right) \pmod{p}.$$

So the quadratic residues are all roots of the congruence

$$x^{\frac{p-1}{2}} \equiv 1 \pmod{p}. \tag{5.19}$$

This polynomial congruence, being of degree $\frac{p-1}{2}$, can have at most $\frac{p-1}{2}$ roots (see section 2.2) so the quadratic residues are all its roots. That means if b is a quadratic non-residue then $b^{\frac{p-1}{2}} \not\equiv 1$ and thus, from (5.18), we have $b^{\frac{p-1}{2}} \equiv -1 = \left(\frac{b}{p}\right)$ (mod p). If r is any non-zero residue, whether it is a quadratic residue or not, we therefore have

$$r^{\frac{p-1}{2}} \equiv \left(\frac{r}{p}\right)(\text{mod } p). \qquad (5.20)$$

This congruence is known as Euler's criterion and it holds even if $p|r$ for then both sides are zero.

Example 5.6

To see whether or not 2 is a square modulo 37, we can now argue that

$$\left(\frac{2}{37}\right) \equiv 2^{\frac{37-1}{2}} = 2^{18}(\text{mod } 37).$$

Successive squaring of powers of 2 then gives $2^2 = 4$, $2^4 = 16$, $2^8 = 256 \equiv -3 \pmod{37}$ and $2^{16} \equiv 9 \pmod{37}$; whence $2^{18} = 2^{16} * 2^2 \equiv 9 * 4 \equiv -1 \pmod{37}$. So 2 is not a square modulo 37.

Activity 5.3 Is 37 irreducible in $\mathbb{Z}[\sqrt{2}]$? (See (5.11) above.)

Example 5.7

We originally deduced the multiplicative property (5.17) of the Legendre symbol directly from the definition. We can also derive it from Euler's criterion. This is because

$$\left(\frac{ab}{p}\right) \equiv (ab)^{\frac{p-1}{2}} = a^{\frac{p-1}{2}} b^{\frac{p-1}{2}} \equiv \left(\frac{a}{p}\right)\left(\frac{b}{p}\right) \pmod{p}.$$

So $\left(\frac{ab}{p}\right)$ and $\left(\frac{a}{p}\right)\left(\frac{b}{p}\right)$ must not only be congruent but equal since they are each either 0, +1 or −1.

⟩5.3 Identifying the rational primes that become composite in a quadratic domain

If, in (5.20), r is −1 we get $\left(\frac{-1}{p}\right) \equiv (-1)^{\frac{p-1}{2}}$ (mod p). Both sides of this congruence must be equal because they are each ±1, so

$$\left(\frac{-1}{p}\right) = (-1)^{\frac{p-1}{2}} = \begin{cases} +1 & \text{if } p \equiv 1 \ (\text{mod } 4) \\ -1 & \text{if } p \equiv -1 \ (\text{mod } 4). \end{cases} \qquad (5.21)$$

Thus -1 is guaranteed to be a quadratic residue if p is a prime congruent to 1 modulo 4 and cannot be a quadratic residue if $p \equiv -1$ (mod 4). This means from (5.11) that

> *an odd prime p will be composite in $\mathbb{Z}[i]$*
>
> *if and only if $p \equiv 1$* (mod 4). $\qquad$ (5.22)

When p is composite in $\mathbb{Z}[i]$, (5.1) shows that there must be a Gaussian integer whose norm is p, and so there are rational integers a and b such that

$$p = (a + bi)(a - bi) = a^2 + b^2.$$

Of course if p is known to be the sum of the two squares a^2 and b^2 then it is automatically $(a + bi)(a - bi)$ and so composite. So

> *an odd prime is a sum of two rational squares if*
>
> *and only if it is congruent to 1 modulo 4.* $\qquad$ (5.23)

The prime 2 is also a sum of two squares, namely $1^2 + 1^2$, corresponding to the factorization $2 = (1 + i)(1 - i)$ in $\mathbb{Z}[i]$.

Activity 5.4 Express each of the primes 5, 13, 29, 41 and 73 as a sum of two squares. Compare your answers with the factorizations of these primes in $\mathbb{Z}[i]$.

When $p = a^2 + b^2 = (a + bi)(a - bi)$ is a sum of two rational integer squares then $a + bi$ and $a - bi$ both have norm p and so are irreducible in $\mathbb{Z}[i]$. We then know that this factorization of p in $\mathbb{Z}[i]$ can only be altered by multiplying the factors by the units 1, -1, i and $-i$. So, apart from order, the number p can only be expressed as

$$(a + bi)(a - bi)$$
$$(-a - bi)(-a + bi)$$
$$(ai - b)(-ai - b)$$
$$\text{or} \qquad (-ai + b)(ai + b).$$

In $\mathbb{Z}$ the corresponding statement is that the only ways p can be expressed as a sum of two squares are as $a^2 + b^2$, $(-a)^2 + (-b)^2$, $a^2 + (-b)^2$ and $(-a)^2 + b^2$.

From (5.22), we see that if q is a prime with $q \equiv -1$ (mod 4), it is irreducible in $\mathbb{Z}[i]$ and its only divisors are the irreducible integers

$$q, \quad -q, \quad qi, \quad -qi. \qquad (5.24)$$

If a prime p is congruent to 1 modulo 4 then in $\mathbb{Z}$ it has a representation $a^2 + b^2$ with $a \neq b$; and in $\mathbb{Z}[i]$ it has the eight irreducible divisors

$$a+bi, \ a-bi, \ -a-bi, \ -a+bi, \ ai-b, \ -ai-b, \ -ai+b, \ ai+b. \quad (5.25)$$

The prime 2 is a sum of two equal squares, so it only has four different prime divisors:

$$1+i, \ 1-i, \ -1+i, \ -1-i. \quad (5.26)$$

Knowing how rational primes factorize enables us to classify all irreducible quadratic integers in $\mathbb{Z}[i]$, because we saw at the beginning of section 5.1 that each irreducible integer occurs as a divisor of some rational prime p. So every irreducible integer in $\mathbb{Z}[i]$ must either be as in (5.24), or arise as one of the prime divisors of 2 or of a rational prime congruent to 1 modulo 4.

In $\mathbb{Q}(\sqrt{d})$ with $d \neq -1$ we could similarly decide which rational primes remain irreducible and which are composite if we could classify all the primes for which d is a quadratic residue. Equation (5.21) gave just such a classification for $d = -1$ and we can get similar classifications for other values of d using more advanced properties of the Legendre symbol. For instance every odd prime must be congruent to one of $1, -1, 3$ or -3 modulo 8, and it is known that the Legendre symbol $\left(\frac{2}{p}\right)$ has the value $+1$ whenever $p \equiv \pm 1 \pmod 8$ and -1 whenever $p \equiv \pm 3 \pmod 8$ (see Appendix 6). So (5.11) means that an odd prime p will be composite in $\mathbb{Z}[\sqrt{2}]$ precisely when $p \equiv \pm 1 \pmod 8$. As in the treatment of $d = -1$, (5.1) shows that when p is composite in $\mathbb{Z}[\sqrt{2}]$, there must be an element of $\mathbb{Z}[\sqrt{2}]$ whose norm is $\pm p$. Actually there must be one whose norm is $+p$, because if the norm of α is $-p$ then the norm of $(1 + \sqrt{2})\alpha$ is $\text{Norm}(1 + \sqrt{2})\text{Norm}(\alpha) = (-1)(-p) = p$. [This is equivalent to observing that if $-p = a^2 - 2b^2$ then we also have $p = (a + 2b)^2 - 2(a + b)^2$.] So when p is composite there will always be rational integers a and b such that

$$p = (a + b\sqrt{2})(a - b\sqrt{2}) = a^2 - 2b^2.$$

Of course if p is of the form $a^2 - 2b^2$ then it is automatically $(a + b\sqrt{2})(a - b\sqrt{2})$ and thus composite. So

an odd prime can be expressed as $a^2 - 2b^2$
if and only if it is congruent to ± 1 modulo 8. \quad (5.27)

In $\mathbb{Z}[\sqrt{2}]$ there are infinitely many units, each of the form $\pm(1 + \sqrt{2})^n$; in particular there are infinitely many units with norm $+1$, each being $\pm(1 + \sqrt{2})^{2k}$ for some $k \in \mathbb{Z}$. So if there is any integer α with norm p there will be infinitely many, all of them multiples of α or its conjugate by the various units of norm $+1$. Each quadratic integer $a + b\sqrt{2}$ of norm p then gives rational integers a, b with $a^2 - 2b^2 = p$. Therefore if a rational prime has any representations in the form $a^2 - 2b^2$, it will have infinitely many such representations. For example 17 can be expressed as $5^2 - 2 * 2^2$ corresponding to the fact that, in $\mathbb{Z}[\sqrt{2}]$, $5 + 2\sqrt{2}$ has norm 17. Then, since $3 + 2\sqrt{2} = (1 + \sqrt{2})^2$ has norm 1, the product $(5 + 2\sqrt{2})(3 + 2\sqrt{2}) = 23 + 16\sqrt{2}$ will also have norm 17, giving $23^2 - 2 * 16^2 = 17$. Again, $(23 + 16\sqrt{2})(3 + 2\sqrt{2}) = 133 + 94\sqrt{2}$ has norm 17, and so $133^2 - 2 * 94^2 = 17$. We can clearly go on producing as many such representations of 17 as we like (as long as we can do the arithmetic!) by continually multiplying integers of norm 17 by $3 + 2\sqrt{2}$ (or by $(3 + 2\sqrt{2})^{-1} = 3 - 2\sqrt{2}$).

Activity 5.5 In $\mathbb{Z}[\sqrt{2}]$ the rational prime $2 = \sqrt{2} * \sqrt{2}$ is composite and the factor $\sqrt{2}$ has norm -2. Multiply $\sqrt{2}$ by $1 + \sqrt{2}$ to form a quadratic integer with norm $+2$, and then find others by repeatedly multiplying by $3 + 2\sqrt{2}$. Thus find at least three different pairs of positive rational integers $(a_1, b_1), (a_2, b_2), (a_3, b_3)$ with $a_i^2 - 2b_i^2 = 2$ for $i = 1, 2, 3$.

Activity 5.6 Which of the rational primes 7, 11, 23, 31, 37, 41 can be expressed in the form $a^2 - 2b^2$? Find suitable values of a and b whenever possible. (You can check your answers by using the computer to see which of them are composite in $\mathbb{Z}[\sqrt{2}]$.)

For an odd prime p we can use our knowledge about $\left(\frac{2}{p}\right)$ and $\left(\frac{-1}{p}\right)$ to find out when $\left(\frac{-2}{p}\right)$ is $+1$ and when it is -1. We again consider the possible residue classes 1, 3, 5, 7 of p modulo 8, and argue that

if $p \equiv 1 \pmod 8$ then $p \equiv 1 \pmod 4$, so that $\left(\frac{2}{p}\right) = +1$, $\left(\frac{-1}{p}\right) = +1$

and $\left(\frac{-2}{p}\right) = \left(\frac{2}{p}\right)\left(\frac{-1}{p}\right) = +1$;

if $p \equiv 3 \pmod 8$ then $p \equiv -1 \pmod 4$, whence $\left(\frac{2}{p}\right) = -1$, $\left(\frac{-1}{p}\right) = -1$ and $\left(\frac{-2}{p}\right) = (-1)(-1) = +1$;

if $p \equiv 5 \pmod 8$ then $p \equiv 1 \pmod 4$, so that $\left(\frac{2}{p}\right) = -1$, $\left(\frac{-1}{p}\right) = +1$

and $\left(\frac{-2}{p}\right) = (-1)(+1) = -1$;

if $p \equiv 7 \pmod 8$ then $p \equiv -1 \pmod 4$, so $\left(\frac{2}{p}\right) = +1$, $\left(\frac{-1}{p}\right) = -1$

and $\left(\frac{-2}{p}\right) = (+1)(-1) = -1$.

Thus -2 is a quadratic residue of those primes that are congruent to either 1 or 3 modulo 8. There is unique factorization in $\mathbb{Z}[\sqrt{-2}]$; so, from (5.11), an odd prime p will be composite in $\mathbb{Z}[\sqrt{-2}]$ precisely when p is in one of these classes. As before, (5.1) then implies that there will be an element in $\mathbb{Z}[\sqrt{-2}]$ of norm $\pm p$ when that happens. Since norms in $\mathbb{Z}[\sqrt{-2}]$ are never negative we can see that if $p \equiv 1 \pmod 8$ or $p \equiv 3 \pmod 8$ there will be an element $a + b\sqrt{-2}$ of norm p. That means

$$an\ odd\ prime\ can\ be\ expressed\ as\ a^2 + 2b^2$$
$$if\ and\ only\ if\ p \equiv 1, 3 \pmod 8. \qquad (5.28)$$

Activity 5.7 Is 2 composite in $\mathbb{Z}[\sqrt{-2}]$?

In $\mathbb{Z}[\sqrt{-2}]$ the factorization corresponding to $p = a^2 + 2b^2$ is $p = (a + b\sqrt{2})(a - b\sqrt{2})$ and we know that this can only be altered by changing the order of the factors or multiplying them by either of the units $+1$ and -1 in $\mathbb{Z}[\sqrt{-2}]$. In $\mathbb{Z}$ this means that if a prime p can be expressed as $a^2 + 2b^2$ for some a, b, then the only other ways of writing p as a square plus twice a square are as $(-a)^2 + 2b^2$, $a^2 + 2(-b)^2$, or $(-a)^2 + 2(-b)^2$ (cf the discussion of primes that are sums of two squares at the beginning of this section).

In finding the values of Legendre symbols with larger numerators, the law of quadratic reciprocity (see Appendix 6) plays a central role. It states that for two different odd primes p and q, the Legendre symbol $\left(\frac{p}{q}\right)$ is equal to $\left(\frac{q}{p}\right)$ if either $p \equiv 1 \pmod 4$ or $q \equiv 1 \pmod 4$, and is equal to $-\left(\frac{q}{p}\right)$ if $p \equiv q \equiv -1 \pmod 4$. The following example shows how we would use it.

Example 5.8
In order to find whether 165 is a square modulo 347 we evaluate the
Legendre symbol $\left(\frac{165}{347}\right)$ as follows.

$$\left(\frac{165}{347}\right) = \left(\frac{3}{347}\right)\left(\frac{5}{347}\right)\left(\frac{11}{347}\right) \quad \text{since } 165 = 3*5*11$$

$$= \left[-\left(\frac{347}{3}\right)\right]*\left(\frac{347}{5}\right)*\left[-\left(\frac{347}{11}\right)\right] \text{ using the law of quadratic}$$

reciprocity

$$= \left(\frac{2}{3}\right)\left(\frac{2}{5}\right)\left(\frac{6}{11}\right)$$

$$= (-1)(-1)\left(\frac{2}{11}\right)\left(\frac{3}{11}\right)$$

$$= \left(\frac{2}{11}\right)\left[-\left(\frac{11}{3}\right)\right] \text{ using quadratic reciprocity again}$$

$$= \left(\frac{11}{3}\right) \text{ since } 11 \equiv 3 \text{ (mod 8) makes } \left(\frac{2}{11}\right) = -1$$

$$= \left(\frac{2}{3}\right) = -1.$$

So 165 is not a square modulo 347.

The computer knows about Legendre symbols and if you choose the
Legendre symbol option from the main menu you can enter any Legendre
symbol or string of symbols, optionally prefaced with a minus sign. You
can then either ask the computer to show you each step that it would take
in simplifying the symbol, or you can tell the computer how you would
do it and it will check that you are doing the simplification correctly.

Activity 5.8 Use the computer's Legendre symbol option to evaluate the
following Legendre symbols.
(i) $\left(\frac{38}{73}\right)$; (ii) $\left(\frac{15}{73}\right)$; (iii) $\left(\frac{239}{587}\right)$; (iv) $\left(\frac{-473}{1987}\right)$; (v) $\left(\frac{783}{1093}\right)$; (vi) $\left(\frac{697}{3511}\right)$.

For any d we could now classify the primes for which d is a quadratic
residue. As with $d = 2$, and -2 above, this would involve considering
the various residue classes that a prime p can lie in, often modulo $4d$, and
working out the value of $\left(\frac{d}{p}\right)$ in each case. For example, when $d = 3$
the law of quadratic reciprocity implies that for $p \neq 2, 3$ the Legendre

symbol $\left(\frac{3}{p}\right)$ will be equal to $\left(\frac{p}{3}\right)$ if $p \equiv 1 \pmod 4$; and will be equal to $-\left(\frac{p}{3}\right)$ if $p \equiv -1 \pmod 4$. Only numbers congruent to 1 modulo 3 are quadratic residues with respect to 3, so we also need to know the residue class of p modulo 3. So we consider the possible residue classes of primes other than 2 or 3 modulo $12 = 3 * 4$. We have

if $p \equiv 1 \pmod{12}$ then $\left(\frac{3}{p}\right) = \left(\frac{p}{3}\right) = \left(\frac{1}{3}\right) = 1$;

if $p \equiv 5 \pmod{12}$ then $\left(\frac{3}{p}\right) = \left(\frac{p}{3}\right) = \left(\frac{2}{3}\right) = -1$;

if $p \equiv 7 \pmod{12}$ then $\left(\frac{3}{p}\right) = -\left(\frac{p}{3}\right) = -\left(\frac{1}{3}\right) = -1$;

if $p \equiv 11 \pmod{12}$ then $\left(\frac{3}{p}\right) = -\left(\frac{p}{3}\right) = -\left(\frac{2}{3}\right) = 1$.

Therefore 3 will be a quadratic residue modulo p if and only if $p \equiv \pm 1$ (mod 12). As before, this implies that p will be composite in $\mathbb{Z}[\sqrt 3]$ when it is in one of these two congruences classes and there will then be an element $a + b\sqrt 3$ in $\mathbb{Z}[\sqrt 3]$ of norm $\pm p$. That is, when $p \equiv \pm 1$ (mod 12), there will be rational integers a, b satisfying

$$a^2 - 3b^2 = \pm p. \tag{5.29}$$

For any given prime p it is easy to see which sign to take in (5.29). This is because $a^2 \equiv 0 \pmod 3$ if $3|a$ and $a^2 \equiv 1 \pmod 3$ if $3{\nmid}a$. If $p \neq 3$ we cannot have $3|a$; so the left side of (5.29) must be congruent to 1 modulo 3. Therefore if $p \equiv 11 \pmod{12}$ we cannot have $a^2 - 3b^2 = p \equiv 2 \pmod 3$ and so in that case we have to have $a^2 - 3b^2 = -p$. Similarly when $p \equiv 1 \pmod{12}$ we cannot have $a^2 - 3b^2 = -p \equiv 2 \pmod 3$ and so must have $a^2 - 3b^2 = p$. For example, 71 is a prime congruent to -1 modulo 12, so from (5.29) we must have either $a^2 - 3b^2 = 71$ or $a^2 - 3b^2 = -71$ for some rational integers a, b. We cannot have $a^2 - 3b^2 = 71 \equiv 2 \pmod 3$, so there must be rational integers a and b such that $a^2 - 3b^2 = -71$; and indeed $2^2 - 3 * 5^2 = -71$.

Activity 5.9 In $\mathbb{Z}[\sqrt 3]$ the integer $2 + 5\sqrt 3$ has norm -71. Multiply repeatedly by the fundamental unit $2 + \sqrt 3$ to produce at least two other quadratic integers of norm -71 and thus find at least two other pairs of rational integers (a_i, b_i) with $a_i^2 - 3b_i^2 = -71$.

Activity 5.10 Find a pair of rational integers a, b with $a^2 - 3b^2 = 13$. Then multiply $a + b\sqrt 3$ by $2 + \sqrt 3$ to produce other quadratic integers of norm 13 and so other pairs of rational integers (a_i, b_i) with $a_i^2 - 3b_i^2 = 13$.

Activity 5.11 Use congruences modulo 3 to show that there are no rational integers a, b with $a^2 - 3b^2 = 2$; and so no quadratic integers in $\mathbb{Z}[\sqrt{3}]$ of norm 2. Use the computer to find (i) some integers in $\mathbb{Z}[\sqrt{3}]$ of norm -2; and (ii) a factorization of 2 in $\mathbb{Z}[\sqrt{3}]$.

Activity 5.12 If there were rational integers a, b with $a^2 - 3b^2 = 3$ then a would be divisible by 3, giving $3a_1^2 - b^2 = 1$ for some a_1. Use congruences modulo 3 to show that this is impossible. Then use the computer to find at least three quadratic integers of norm -3.

The value of $\left(\frac{-d}{p}\right)$ is always related to that of $\left(\frac{d}{p}\right)$ by $\left(\frac{-d}{p}\right) = \left(\frac{-1}{p}\right)\left(\frac{d}{p}\right)$. So once we have decided which values $\left(\frac{d}{p}\right)$ may take according to the residue classes that p may lie in, we can immediately write down the possible values of $\left(\frac{-d}{p}\right)$, since they will be the same if $p \equiv 1 \pmod 4$ and opposite in sign if $p \equiv -1 \pmod 4$. For example, from the fact that

$$\left(\frac{3}{p}\right) = \begin{cases} +1 & \text{if } p \equiv \pm 1 \pmod{12} \\ -1 & \text{if } p \equiv \pm 5 \pmod{12} \end{cases} \tag{5.30}$$

we can immediately deduce that

$$\left(\frac{-3}{p}\right) = \begin{cases} +1 & \text{if } p \equiv 1 \text{ or } 7 \pmod{12} \\ -1 & \text{if } p \equiv 5 \text{ or } 11 \pmod{12} \end{cases} \tag{5.31}$$

because if $p \equiv 1$ or 5 (mod 12) it is also 1 modulo 4, meaning that $\left(\frac{-3}{p}\right) = \left(\frac{3}{p}\right)$ in these cases; and if $p \equiv -1$ or -5 (mod 12) it is also -1 modulo 4, meaning that $\left(\frac{-3}{p}\right) = -\left(\frac{3}{p}\right)$ in these cases. Notice that primes congruent to either 1 or 7 modulo 12 are precisely those congruent to 1 modulo 3; and primes congruent to 5 or 11 modulo 12 are the odd primes congruent to 2 modulo 3. So we can condense (5.31) to

$$\left(\frac{-3}{p}\right) = \begin{cases} +1 & \text{if } p \equiv 1 \pmod 3 \\ -1 & \text{if } p \equiv 2 \pmod 3. \end{cases} \tag{5.32}$$

The uniqueness of factorization in $\mathbb{Q}(\sqrt{-3})$ now means that a prime p will be composite in that domain if $p \equiv 1 \pmod 3$ and will be irreducible if $p \equiv 2 \pmod 3$. [We have already seen half of this result in the remark just after (5.5).] So if $p \equiv 1 \pmod 3$ there will be an element $\frac{a+b\sqrt{-3}}{2}$ in $\mathbb{Q}(\sqrt{-3})$ with $\text{Norm}\left(\frac{a+b\sqrt{-3}}{2}\right) = \pm p$. Since norms are positive in this domain there will be rational integers a and b with $a \equiv b \pmod 2$ and

$$\frac{a^2 + 3b^2}{4} = p$$

or

$$a^2 + 3b^2 = 4p. \tag{5.33}$$

If a and b are both even here then we have $\left(\frac{a}{2}\right)^2 + 3\left(\frac{b}{2}\right)^2 = p$. If a and b are both odd then they each have to be congruent to 1 or 3 modulo 4. This implies that either $a \equiv b \pmod 4$ or $a \equiv -b \pmod 4$. In the first case $\frac{a+3b}{4}$ and $\frac{a-b}{4}$ are rational integers; and in the second case $\frac{a-3b}{4}$ and $\frac{a+b}{4}$ are rational integers. In either case notice that, from (5.33),

$$\left(\frac{a \pm 3b}{4}\right)^2 + 3\left(\frac{a \mp b}{4}\right)^2 = \frac{a^2 + 3b^2}{4} = p. \tag{5.34}$$

So, $a^2 + 3b^2 = 4p$ with $a \equiv b \pmod 2$ always implies the existence of rational integers x, y such that

$$x^2 + 3y^2 = p. \tag{5.35}$$

Of course, given (5.35), we would immediately have $(2x)^2 + 3(2y)^2 = 4p$ with $2x \equiv 2y \pmod 2$. Being able to write p in the form $x^2 + 3y^2$ with integral x, y is therefore completely equivalent to being able to express $4p$ as $a^2 + 3b^2$ with $a \equiv b \pmod 2$. If $p \equiv 1 \pmod 3$ we have both (5.33) and (5.35), and if $p \equiv 2 \pmod 3$ we have neither. This statement remains true if $p = 2$ for neither (5.33) nor (5.35) can be satisfied then and, correspondingly, the prime 2 is irreducible in $\mathbb{Q}(\sqrt{-3})$. The prime 3 also has to be treated separately, but obviously $3 = 0^2 + 3 * 1^2$ and $3 = \sqrt{-3} * \sqrt{-3}$ is composite in $\mathbb{Q}(\sqrt{-3})$.

Another way of looking at the connection between (5.33) and (5.35) is that if there is a quadratic integer $\frac{1}{2}(a + b\sqrt{-3})$ of norm p, there must also be one of the form $x + y\sqrt{-3}$ where the rational coefficients of 1 and $\sqrt{-3}$ are rational integers and not just halves of rational integers. This comes about because there are six units in the domain of integers in $\mathbb{Q}(\sqrt{-3})$. So any factorization of a rational prime in $\mathbb{Q}(\sqrt{-3})$ gives rise to six factorizations and some of the resulting factors will have rational integer coefficients of 1 and $\sqrt{-3}$.

Example 5.9

The rational prime 13 is the norm of $\frac{5+3\sqrt{-3}}{2}$, corresponding to the factorization $\left(\frac{5+3\sqrt{-3}}{2}\right) * \left(\frac{5-3\sqrt{-3}}{2}\right) = 13$. We can alter this by multiplying one factor by any of the six units $1, -1, \frac{1+\sqrt{-3}}{2}, \frac{1-\sqrt{-3}}{2}, \frac{-1+\sqrt{-3}}{2}$ or $\frac{-1-\sqrt{-3}}{2}$ and the other factor by the inverse unit. The result is the six factorizations:

$$\left(\frac{5+3\sqrt{-3}}{2}\right)\left(\frac{5-3\sqrt{-3}}{2}\right)$$

$$\left(\frac{-5-3\sqrt{-3}}{2}\right)\left(\frac{-5+3\sqrt{-3}}{2}\right)$$

$$\left(\frac{-2+4\sqrt{-3}}{2}\right)\left(\frac{-2-4\sqrt{-3}}{2}\right)$$

$$\left(\frac{7-\sqrt{-3}}{2}\right)\left(\frac{7+\sqrt{-3}}{2}\right)$$

$$\left(\frac{-7+\sqrt{-3}}{2}\right)\left(\frac{-7-\sqrt{-3}}{2}\right)$$

$$\left(\frac{2-4\sqrt{-3}}{2}\right)\left(\frac{2+4\sqrt{-3}}{2}\right)$$

So the third and the last of these show that 13 actually has the factorizations $(-1+2\sqrt{-3})(-1-2\sqrt{-3})$ and $(1-2\sqrt{-3})(1+2\sqrt{-3})$ with integer coefficients of 1 and $\sqrt{-3}$.

Activity 5.13 The rational prime 31 has the factorization $\left(\frac{7+5\sqrt{-3}}{2}\right)$ $* \left(\frac{7-5\sqrt{-3}}{2}\right)$ in $\mathbb{Q}(\sqrt{-3})$. For each unit $\eta \in \mathbb{Q}(\sqrt{-3})$, calculate $\left(\frac{7+5\sqrt{-3}}{2}\right)\eta$ and $\left(\frac{7-5\sqrt{-3}}{2}\right)/\eta$ and thus find the other five factorizations of 31 in $\mathbb{Q}(\sqrt{-3})$.

It is straightforward to classify the primes for which 5 is a square because for $p \neq 2, 5$ we have

$$\left(\frac{5}{p}\right) = \left(\frac{p}{5}\right) \text{ using quadratic reciprocity}$$

$$= \begin{cases} +1 & \text{if } p \equiv \pm 1 \pmod 5 \\ -1 & \text{if } p \equiv \pm 2 \pmod 5. \end{cases} \tag{5.36}$$

Again this means that the odd primes congruent to ± 1 modulo 5 are composite in $\mathbb{Q}(\sqrt{5})$ and those congruent to ± 2 are irreducible. The prime 2 is irreducible (see Activity 5.2) and $5 = \sqrt{5} * \sqrt{5}$ is of course composite. All the composite rational primes satisfy

$$\frac{a^2 - 5b^2}{4} = \pm p \text{ with } a \equiv b \pmod 2 \tag{5.37}$$

and we can always take the plus sign here since if α has norm $-p$ then $\left(\frac{1+\sqrt{5}}{2}\right)\alpha$ has norm p. Notice that if $\frac{a+b\sqrt{5}}{2}$ has norm p and a, b are both odd then $a \equiv b \pmod 4$ or $a \equiv -b \pmod 4$. In the former case

$$\left(\frac{3+\sqrt{5}}{2}\right)\left(\frac{a+b\sqrt{5}}{2}\right) = \frac{(3a+5b)+(3b+a)\sqrt{5}}{4}$$

has norm p and integer coefficients of 1 and $\sqrt{5}$, and in the latter case

$$\left(\frac{3-\sqrt{5}}{2}\right)\left(\frac{a+b\sqrt{5}}{2}\right) = \frac{(3a-5b)+(3b-a)\sqrt{5}}{4}$$

has norm p and integer coefficients for 1 and $\sqrt{5}$. So for 5 and every prime $\equiv \pm 1 \pmod 5$ we can write

$$p = x^2 - 5y^2 \text{ for some integers } x, y. \tag{5.38}$$

Activity 5.14 Express each of 5, 11, 19 in the form $x^2 - 5y^2$ with rational integer values of x and y.

As in previous cases it is straightforward to deduce that

$$\left(\frac{-5}{p}\right) = \begin{cases} +1 & \text{if } p \equiv 1, 3, 7, 9 \pmod{20} \\ -1 & \text{if } p \equiv 11, 13, 17, 19 \pmod{20}. \end{cases} \tag{5.39}$$

This means that the only rational primes which remain prime in $\mathbb{Z}[\sqrt{-5}]$ are those congruent to 11, 13, 17 or 19 modulo 20. However we cannot use (5.39) alone to decide which rational primes are irreducible in $\mathbb{Z}[\sqrt{-5}]$ since we do not have uniqueness of factorization here. We have to use (5.1) and decide which rational primes can be norms in this quadratic domain. All norms in $\mathbb{Z}[\sqrt{-5}]$ are positive and of the form $a^2 + 5b^2$ but it is not immediately clear which primes can be represented in this form for suitable rational integers a and b. Certainly (5.39) implies that the equation

$$p = a^2 + 5b^2 \qquad (5.40)$$

cannot have solutions if $p \equiv 11, 13, 17, 19 \pmod{20}$. Also we saw in Activity 5.2 that it has no solutions if $p \equiv 2$ or $3 \pmod 5$. So the only primes that could be of the form $a^2 + 5b^2$ are 5 and those congruent to 1 or 9 modulo 20.

Example 5.10
Neither 2 nor 3 are norms in $\mathbb{Z}[\sqrt{-5}]$, so they are each irreducible quadratic integers in this domain. Neither of them is a prime quadratic integer though, because 6 has the two factorizations

$$2 * 3 = (1 + \sqrt{-5})(1 - \sqrt{-5}).$$

Here 2 divides the right hand product but does not divide $1 + \sqrt{-5}$ or $1 - \sqrt{-5}$ as Norm(2) = 4 does not divide Norm$(1 + \sqrt{-5})$ = 6 or Norm$(1 - \sqrt{-5})$ = 6. Similarly 3 does not divide either $1 + \sqrt{-5}$ or $1 - \sqrt{-5}$.

Obviously $5 = 0^2 + 5 * 1^2$; and it is also true that every prime congruent to 1 or 9 modulo 20 can be expressed as $a^2 + 5b^2$. The proof of this is not difficult, but requires some background knowledge about quadratic forms (see e.g. chapters 1 and 2 of Watson (1960)). So, as regards factorization in $\mathbb{Z}[\sqrt{-5}]$, rational primes fall into three types:

those congruent to 11, 13, 17, 19 modulo 20 are prime in $\mathbb{Z}[\sqrt{-5}]$;

those congruent to 2, 3 or 7 modulo 20 are irreducible

but not prime in $\mathbb{Z}[\sqrt{-5}]$;   (5.41)

those congruent to 1, 5 or 9 modulo 20 are composite in $\mathbb{Z}[\sqrt{-5}]$.

Activity 5.15 Express 29 and 41 in the form $a^2 + 5b^2$ and thus factor them in $\mathbb{Z}[\sqrt{-5}]$.

⟩5.4 Sums of squares

We can use our knowledge of which primes are sums of two squares to classify all the numbers that are sums of two squares. This is because there is a classical identity showing that the product of two numbers that are each sums of two squares is again a sum of two squares: namely

$$(a^2 + b^2)(x^2 + y^2) = (ax - by)^2 + (ay + bx)^2. \qquad (5.42)$$

This is the same as the fact that the norm in $\mathbb{Z}[i]$ is multiplicative; because the left-hand side of (5.42) is $\text{Norm}(a + bi)*\text{Norm}(x + yi)$ and, as $(a + bi)(x + yi) = (ax - by) + (ay + bx)i$, the right-hand side is $\text{Norm}[(a + bi)(x + yi)]$.

Example 5.11
From $5 = 1^2 + 2^2$ and $13 = 2^2 + 3^2$, (5.42) gives

$$65 = (1^2 + 2^2)(2^2 + 3^2) = (1*2 - 2*3)^2 + (1*3 + 2*2)^2 = 4^2 + 7^2.$$

Note that changing the sign of b would alter (5.42) to $(a^2 + (-b)^2)(x^2 + y^2) = (ax + by)^2 + (ay - bx)^2$; which in this instance gives

$$65 = (1*2 + 2*3)^2 + (1*3 - 2*2)^2 = 8^2 + 1^2.$$

When (5.42) is used with (5.23) it immediately implies that any product composed of powers of 2 and of primes congruent to 1 modulo 4 is a sum of two squares. Also, although no prime $q \equiv -1 \pmod{4}$ can be a sum of two squares, any even power of q is such a sum, since $q^{2s} = (q^s)^2 + 0^2$. So any natural number is a sum of two squares if it is

> *a product of: powers of 2; powers of primes congruent*
> *to 1 modulo 4 and even powers of primes congruent*
> *to −1 modulo 4.* (5.43)

We can be sure that this is the complete description of natural numbers of the form $a^2 + b^2$ if we show that a number that is a sum of two squares and is divisible by a prime $q \equiv -1 \pmod{4}$ must be exactly divisible by an even power of q. This is certainly true for the first few numbers 1, 2, 4, 5, 8, 9, ... that are sums of two squares. Suppose then that n is a sum of two squares, that it is divisible by a prime $q \equiv -1 \pmod{4}$, and that all sums of two squares less than n are of the form (5.43). So

$$n = x^2 + y^2 \equiv 0 \pmod{q}. \qquad (5.44)$$

If y were not divisible by q it would have an inverse, say z, modulo q; leading to $x^2 \equiv -y^2$ (mod q) and $(xz)^2 \equiv -1$ (mod q). But that would be impossible as -1 is not a square modulo q. So in (5.44), y, and so x, must be divisible by q and thus

$$\frac{n}{q^2} = \left(\frac{x}{q}\right)^2 + \left(\frac{y}{q}\right)^2.$$

Now $\frac{n}{q^2} < n$, whence by assumption it is of the form (5.43). Hence $\frac{n}{q^2}$ and so also n, are only divisible by even powers of q.

Activity 5.16 Which of the following natural numbers are sums of two squares? (i) 306; (ii) 1411; (iii) 221; (iv) 496; (v) 1093. Later, the sums of squares computer program will provide an easy way to check your answers.

The identity (5.42) can be generalized to

$$(a^2 - db^2)(x^2 - dy^2) = (ax + dby)^2 - d(ay + bx)^2 \qquad (5.45)$$

because in $\mathbb{Q}(\sqrt{d})$ the left-hand side is the product of $\text{Norm}(a + b\sqrt{d})$ and $\text{Norm}(x + y\sqrt{d})$ and the right-hand side is $\text{Norm}[(ax + dby) + (ay + bx)\sqrt{d}] = \text{Norm}[(a + b\sqrt{d})(x + y\sqrt{d})]$. Just as (5.42) led to the complete classification of numbers that are sums of two squares, so (5.45) enables us to generalize some of our earlier results about representation of primes by $a^2 - db^2$. For example we know from (5.28) and Activity 5.7 that the only primes that can be written in the form $a^2 + 2b^2$ are 2 and odd primes congruent to 1 or 3 modulo 8. So, from (5.45) with $d = -2$, any natural number that is a product of (any number of) primes congruent to 1, 2, 3 modulo 8 will also be of the form $a^2 + 2b^2$. Also, when $q \equiv 5$ or 7 (mod 8) the congruence $x^2 + 2y^2 \equiv 0$ (mod q) is only possible with $q|x$ and $q|y$ (cf (5.44)). So, as in the argument after (5.43), this leads to the complete classification of natural numbers of the form $a^2 + 2b^2$. They must be

> *products of: powers of 2; powers of primes congruent to 1 or 3 modulo; 8 and even powers of primes congruent to 5 or 7 modulo 8.* $\qquad (5.46)$

Activity 5.17 Express 17 and 41 in the form $a^2 + 2b^2$ for rational integers a, b and then use the case $d = -2$ of (5.45) to deduce a representation of $17 * 41 = 697$ in the same form.

In Example 5.11 we got a different identity for products of sums of squares by changing the sign of one of the variables. Similarly, altering the sign of b in (5.45) gives

$$(a^2 - db^2)(x^2 - dy^2) = (ax - dby)^2 - d(ay - bx)^2. \qquad (5.47)$$

For example from $11 = 3^2 + 2 * 1^2$ and $19 = 1^2 + 2 * 3^2$, (5.45) gives

$$209 = 11 * 19 = (3 * 1 - 2 * 1 * 3)^2 + 2(3 * 3 + 1 * 1)^2 = (-3)^2 + 2 * 10^2$$

and (5.47) gives

$$209 = (3 * 1 + 2 * 1 * 3)^2 + 2(3 * 3 - 1 * 1)^2 = 9^2 + 2 * 8^2.$$

The two identities (5.45) and (5.47) do not always lead to completely different representations. In the case of $d = -1$ for instance, (5.45) applied to the number $90 = (3^2 + 3^2)(2^2 + 1^2)$ gives $(3*2 - 3*1)^2 + (3*1 + 3*2)^2 = 3^2 + 9^2$; whereas (5.47) gives $90 = (3*2 + 3*1)^2 + (3*1 - 3*2)^2 = 9^2 + (-3)^2$. These two representations are essentially the same, only differing in the order and sign of the terms. Representations that are either the same or differ trivially are always produced when any one of a, b, x, y is zero; or, in the case of $d = -1$, when $a = \pm b$ or $x = \pm y$.

Activity 5.18 Take the representations of 17 and 41 that you found in Activity 5.17 and combine them by using the case $d = -2$ of (5.47). Do you get a different representation of 697 as a square plus twice a square?

Example 5.12
From (5.35) and the following paragraph, the case $(a^2 + 3b^2)(x^2 + 3y^2) = (ax - 3by)^2 + 3(ay + bx)^2$ of (5.45) implies that every natural number that is a product of primes congruent to 1 modulo 3 is of the form $a^2 + 3b^2$. Also $q^{2s} = (q^s)^2 + 3 * 0^2$ for any $q \equiv -1$ (mod 3); so as before,

> *natural numbers of the form $a^2 + 3b^2$ are products of*
> *powers of 3; powers of primes congruent to 1 modulo 3;*
> *and even powers of primes congruent to 2 modulo 3.*   (5.48)

The generalizations of (5.27) and (5.38) in the cases $d = 2, 5$ are very similar. They are

> *rational integers of the form $a^2 - 2b^2$ are products of: ± 1;*
> *powers of 2; powers of primes congruent to ± 1 modulo 8;*
> *and even powers of primes congruent to ± 3 modulo 8.*   (5.49)

and

> *integers of the form $a^2 - 5b^2$ are products of:* ± 1;
>
> *powers of 5; powers of primes congruent to* ± 1 *modulo 5;*
>
> *and even powers of primes congruent to* ± 2 *modulo 5.*   (5.50)

The generalization of (5.29) in the case $d = 3$ is also straightforward, but because of the $\pm$ sign in (5.29) we have to express the result as

> *a rational integer n will be of the form $a^2 - 3b^2$*
>
> *if and only if it is 1 modulo 3 and is a product of*
>
> ± 1; *powers of 2; powers of primes congruent to* ± 1 *modulo 12;*
>
> *and even powers of primes congruent to* ± 5 *modulo 12.*   (5.51)

The case $d = -5$ is not as simple as the previous cases because, as we know, in $\mathbb{Z}[\sqrt{-5}]$ there are rational primes that are irreducible but not prime in the quadratic domain. That is, there are rational primes such as 3 and 7 that are not norms in $\mathbb{Z}[\sqrt{-5}]$ but can divide norms of irreducible integers. We can see the difference this makes if we follow the previous analysis. Here the identity (5.45) becomes

$$(a^2 + 5b^2)(x^2 + 5y^2) = (ax - 5by)^2 + 5(ay + bx)^2 \qquad (5.52)$$

and this guarantees that any natural number that is a product of primes, each of the form $a^2 + 5b^2$, will be of the same form. For example $5 = 0^2 + 5 * 1^2$ and $29 = 3^2 + 5 * 2^2$ so, from (5.52), $145 = 5 * 29 = (0 - 5 * 1 * 2)^2 + 5 * (0 + 1 * 3)^2 = 10^2 + 5 * 3^2$. So (5.52), used with (5.41), implies that any product of powers of 5 and powers of primes congruent to 1 or 9 modulo 20 will be of the form $a^2 + 5b^2$ for some rational integers a and b. However that does not describe all the natural numbers of this form: $21 = 3 * 7$ for instance is not such a product but nonetheless $21 = 1^2 + 5 * 4^2$. This is typical of the situation where there is not unique factorization among the integers of $\mathbb{Q}(\sqrt{d})$. The identity (5.45) always implies that any number that is composed of rational primes that are norms in $\mathbb{Q}(\sqrt{d})$ will again be of the form $a^2 - db^2$ for rational integers a, b. But if there are rational primes that are irreducible but not prime in $\mathbb{Q}(\sqrt{d})$ there will be rational integers, like 21 in the case of $\mathbb{Q}(\sqrt{-5})$, that are norms but are not products of norms. The whole story in these cases of non-unique factorization involves the 'composition theory' of quadratic forms and will not be treated here (see Cohn(1980)).

The *sums of squares* computer program enables you to explore some of the many ways of writing numbers as sums of squares. The program restricts itself to trying to express natural numbers as sums of squares with positive coefficients and concentrates on the forms $x^2 + y^2$, $x^2 + 2y^2$, $x^2 + 3y^2$ and $x^2 + 5y^2$ that we investigated above.

Example 5.13
Choose the sums of squares option from the computer's main menu and enter the number 28665. As soon as you press ENTER the computer factorizes the number and displays $21^2 * 5 * 13$ in this case. Here the factors shown are not all prime, because when the number is a sum of two squares, the computer amalgamates any primes congruent to -1 modulo 4 into a common squared factor. Just pressing ENTER now produces $21^2(2^2 + 1^2)(3^2 + 2^2)$ and then ENTER again makes the computer use (5.42) to give $84^2 + 147^2$.

Activity 5.19 Use the sums of squares option to express each of the following numbers as a sum of two squares: (i) 4985; (ii) 10660; (iii) 230114; (iv) 255476; (v) 193738 (vi) 310298.

Activity 5.20 Press the function key F3 to make the computer consider sums of the form $x^2 + 3y^2$ and then use the computer to write each of the following numbers in that way: (i) 127; (ii) 247; (iii) 78921; (iv) 38476; (v) 659113; (vi) 900025.

Example 5.14
Pressing the 'CTRL' and '→' keys simultaneously enables us to add the squares of numbers we have chosen. Thus it is easy to see that $6^2 = 36$, $67^2 = 4489$ and $678^2 = 459684$. Pressing F1 if necessary to ensure that we are working with $x^2 + y^2$, it is also straightforward that $678^2 + 426^2 = 641160$, and similarly that $876^2 + 124^2 = 782752$. Are these sums the same as those produced when we ask the computer to automatically express 641160 and 782752 as sums of squares?

Suppose we have asked the computer to investigate numbers of one of the forms $x^2 - dy^2$ for $d = -1, -2, -3, -5$ and that we have entered a number of this form which has two or more factors of the same form. The computer will always combine these factors by using the identity (5.45). However, since it always shows the representations of the factors, you can combine two or more of the factors using (5.47) if you prefer, and check your representation by using the computer to add your squares.

Activity 5.21 Repeat Activity 5.20 but, if possible, find new representations of the given numbers in the form $x^2 + 3y^2$ by using the case $d = -3$ of (5.47) whenever there are two or more factors.

When the computer is automatically trying to represent numbers as sums of squares and we enter a number that is not of one of the forms $x^2 + y^2, \ldots, x^2 + 5y^2$ the computer will then include more terms and see whether the number can be written as sums such as $x^2 + y^2 + z^2$ or $x^2 + 2y^2 + z^2$. For instance suppose that we are working with $x^2 + y^2$ and enter the number 86. Its factorization, $2 * 43$, should tell us that it cannot be a sum of two squares since one of the factors is congruent to -1 modulo 4 (see (5.43)); and pressing ENTER again confirms that it is not. When we press ENTER after that the computer then gives $86 = 7^2 + 6^2 + 1^2$. Even three terms are not enough sometimes. The number 87 for instance cannot be written as $x^2 + y^2 + z^2$ with rational integers x, y, z; and 58 cannot be written as $x^2 + 2y^2 + z^2$. The computer never has to include more than four terms though, because

$$\textit{for each k=1, 2, 3, 5, every natural number can} \qquad (5.53)$$

$$\textit{be written as } x^2 + ky^2 + z^2 + t^2$$

for some rational integers x, y, z, t. The case $k = 1$ is due to Lagrange in 1770 and says that *every natural number is a sum of four squares of rational integers* (see e.g. Jackson(1987)). In 1917 Ramanujan generalized Lagrange's result and investigated all the sets of positive integer coefficients a, b, c, d, each of which has the property that every natural number can be written in the form $ax^2 + by^2 + cz^2 + dt^2$ for rational integers x, y, z, t. Dickson(1927) later corrected an oversight in Ramanujan's work, so it is now known that there are exactly 54 suitable sets of coefficients a, b, c, d. The complete list of possible coefficients is in Ramanujan(1917) and includes the cases in (5.53). Many other intriguing results are known about sums of squares (see e.g. Cox(1989) and Grosswald(1985)) and lead on to other beautiful theories.

) SUMMARY OF CHAPTER 5

One of the central questions examined in this chapter was 'when does a rational prime p become composite in a quadratic domain?' This happens precisely when there is a quadratic integer whose norm is $\pm p$ and that in turn gives an equation in rational integers that must be satisfied in order for p to be composite. Sometimes congruence considerations immediately show that p is irreducible; and in any case they lead to further criteria. The case $p = 2$ is usually easy to deal with, as is the case where there is unique factorization among the integers of $\mathbb{Q}(\sqrt{d})$. If there is unique factorization in $\mathbb{Q}(\sqrt{d})$ then an odd rational prime p will be a composite integer in $\mathbb{Q}(\sqrt{d})$ if and only if d is a square modulo p; conversely p will be a prime quadratic integer if and only if d is not a square modulo p. If we do not have uniqueness of factorization then it is still true that if d is not a square modulo p then p will remain prime in $\mathbb{Q}(\sqrt{d})$; but if d is a square then p may be either composite or irreducible in $\mathbb{Q}(\sqrt{d})$, though it definitely will not possess the prime property.

The question of when d is a square modulo p can be tackled entirely within $\mathbb{Z}$ with the help of the Legendre symbol $\left(\frac{d}{p}\right)$. This symbol can be manipulated with the aid of simple rules that enable its value to be calculated easily (see section 5.2 and Appendix 6). So we can tell quickly whether or not d is a square modulo p without searching for a squareroot. Not only can we evaluate $\left(\frac{d}{p}\right)$ for a particular d and p but the rules let us find all the primes p which make d a square modulo p. This means that when there is unique factorization among the integers of $\mathbb{Q}(\sqrt{d})$ we can classify all the rational primes that become composite in that domain. Thus in section 5.3 we proved that in $\mathbb{Z}[i]$ the only odd primes that are composite are those congruent to 1 modulo 4. The corresponding statement in $\mathbb{Z}$ is that such primes are the only odd primes that are sums of two rational integer squares. We similarly classified all the rational primes that become composite in the domain of integers of $\mathbb{Q}(\sqrt{d})$ for $d = -2, 2, -3, 3$ and 5. In each case the same rational primes are those represented by $x^2 - dy^2$ for integers x, y and they can be distinguished by congruence conditions. When $d = -5$ the situation is rather more subtle because $\mathbb{Z}[\sqrt{-5}]$, the domain of integers in $\mathbb{Q}(\sqrt{-5})$, does not have uniqueness of factorization. So in $\mathbb{Z}[\sqrt{-5}]$ the rational primes fall into three types: those that are still prime as quadratic integers; those that are irreducible but not prime; and those that are composite in $\mathbb{Z}[\sqrt{-5}]$. These latter are the primes represented in $\mathbb{Z}$ by the form $x^2 + 5y^2$; and here again the different types lie in different congruence classes though that is harder to prove.

   The last section used the multiplicative nature of the norm in $\mathbb{Q}(\sqrt{d})$ to deduce identities that let us classify all the rational integers represented in $\mathbb{Z}$ by $x^2 - dy^2$ for $d = -3, -2, -1, 2, 3, 5$. The classification of all rational integers represented by $x^2 + 5y^2$ is not as straightforward and involves more advanced ideas. However the computer makes it easy to explore ways of writing given numbers in one of the shapes $x^2 + y^2$, $x^2 + 2y^2$, $x^2 + 3y^2$, $x^2 + 5y^2$. It will also help to generate values of these forms.

⟩ EXERCISES FOR CHAPTER 5

1. If p is an odd prime and a is any rational integer show that the congruence $x^2 \equiv a \pmod{p}$ has exactly $1 + \left(\frac{a}{p}\right)$ solutions, where $\left(\frac{a}{p}\right)$ is a Legendre symbol.

2. Determine the number of solutions in each of these cases without trying to solve the congruences. (i) $x^2 \equiv 2 \pmod{17}$; (ii) $x^2 \equiv 8 \pmod{17}$; (iii) $x^2 - 6x + 1 \equiv 0 \pmod{17}$ [complete the square]; (iv) $x^2 \equiv 2 \pmod{101}$; (v) $x^2 \equiv -1 \pmod{61}$; (vi) $x^2 \equiv -3 \pmod{61}$.

3. Show that $x^4 + 1$ is irreducible in $\mathbb{Z}[x]$ but has non-trivial factors in $\mathbb{Z}_p[x]$ for every p. [Hint: $x^4 + 1$ can also be written as $(x^2 + 1)^2 - 2x^2$ and as $(x^2 - 1)^2 + 2x^2$. Can any of these alternatives be a square or a difference of two squares in $\mathbb{Z}_p[x]$?]

4. Prove that $\sqrt{91}$ cannot be the hypotenuse of a right-angled triangle with rational sides.

5. Suppose the square-free integer d is congruent to 1 modulo 4 and that $\sigma = \frac{1+\sqrt{d}}{2}$. If r and s are rational integers and $\alpha = r + s\sigma$ show that $\mathrm{Norm}(\alpha) = r^2 + rs + \left(\frac{1-d}{4}\right)s^2$. By choosing d appropriately deduce that a prime natural number p can only be written in the form $r^2 + rs + s^2$ if $p = 3$ or $p \equiv 1 \pmod 3$.

6. Show that each quadratic integer in $\mathbb{Q}(\sqrt{-3})$ must have an associate of the form $u + v\sqrt{-3}$ where u, v are rational integers. Is a similar statement true about quadratic integers in $\mathbb{Q}(\sqrt{5})$?

7. Verify that the identities (5.45) and (5.47) give the same or trivially different representations provided that one of a, b, x, y is zero, or, in the case $d = -1$, when $a = \pm b$ or $x = \pm y$.

8. Show that there cannot be an identity like (5.42) for sums of three squares by finding an example of two natural numbers that are each sums of three squares but whose product is not such a sum.

9. Use congruences modulo 8 to show that a rational integer $8k + 7$ cannot be the sum of three squares of rational integers.

10. Use congruences modulo 4 to prove that the equation $x^2 + y^2 + z^2 = 4m$ can only hold if each of x, y, z is even. Deduce from the last question that no rational integer of the form $4^n(8k + 7)$ can be a sum of three squares of rational integers.

11. Follow the method of the previous question to prove that no rational integer of the form $4^n(8k + 3)$ can be expressed as $x^2 + y^2 + 5z^2$ for x, y, z rational integers.

〉 Appendix 1

〉 Abstract perspectives

〉1.1 Groups

The simplest algebraic structure that we shall need is a *group*. This is a set together with a special kind of rule or operation for combining any two of its elements. If the set is called G and the rule is denoted by $\star$, then the structure is a group if the rule satisfies the following properties or *axioms*.

(i) *Closure property*: if a and b are in G, then $a \star b$, the result of combining them, must also be in G;

(ii) *Associative property*: if a, b and c are in G, then $a \star (b \star c) = (a \star b) \star c$;

(iii) there is a special element $e \in G$, called the *identity* of G, such that

$$a \star e = e \star a = a \text{ for any } a \in G;$$

(iv) for each element $a \in G$ there is a corresponding element $a' \in G$, the *inverse* of a, such that

$$a \star a' = a' \star a = e.$$

We usually speak of the group G rather than, more accurately, the group $(G, \star)$ and often use the shorthand '$a \in G$' to mean that a is an element of G. Common examples of groups are the set of rational numbers (or the set of real numbers or the set of complex numbers) with the operation of addition; the set of non-zero rationals (or non-zero real numbers, or non-zero complex numbers) with the operation of multiplication; the set of integers with the operation of addition, but *not* the set of non-zero integers with the operation of multiplication (since 2 for example would not have an inverse in the same set). In all these examples the result of combining two elements a and b is independent of the order in which the elements are considered: $2 + 4 = 4 + 2$ and $\frac{3}{7} * \frac{4}{11} = \frac{4}{11} * \frac{3}{7}$ for instance. If, in a group $(G, \star)$, the rule $\star$ always satisfies

(v) *Commutative property*: if a and b are in G then $a \star b = b \star a$

the group is said to be *Abelian* (after N. H. Abel, 1802–1829). All the groups we shall encounter will be Abelian.

The groups mentioned above are *infinite groups* as they each contain an infinite number of elements. For each integer $m > 1$ we can easily construct a *finite group* containing m elements if we first define a new operation of 'addition modulo m' on the integers in the range $0, \ldots, m-1$. We add two of these numbers a and b as usual and take their sum modulo m to be the number s, between 0 and $m - 1$, that satisfies $s \equiv a + b$ (mod m). The set of integers $0, 1, \ldots, m-1$ with this operation is then a group. Another example is the set of integers less than m, and relatively prime to it, with the analogously defined operation of 'multiplication modulo m' where we first multiply a and b as usual and then define their product modulo m to be the number between 0 and $m - 1$ that is congruent to ab (mod m). In this example the element a has an inverse because it is always possible to find an integer a', in the same set, with $aa' \equiv 1$ (mod m), see Jackson(1987).

In any group we can define powers of an element a in the group by writing a^k as shorthand for $a \star a \star \ldots \star a$, where there are k terms in this 'product'. In a finite group with n elements the particular power a^n has a simple value independent of a. In order to see this we suppose that the elements of the group are $x_1, x_2, \ldots, x_n$, (where a is one of these) and consider all the terms $a \star x_1, a \star x_2, \ldots, a \star x_n$. These 'multiples' of a are all distinct, because if $a \star x_i = a \star x_j$ we would have $a' \star (a \star x_i) = a' \star (a \star x_j)$, and so $(a' \star a) \star x_i = (a' \star a) \star x_j$, immediately leading to $x_i = ex_i = ex_j = x_j$. Since there are n multiples, which are each in the group, and n elements altogether, the n multiples $a \star x_1, a \star x_2, \ldots, a \star x_n$ must be the n elements $x_1, x_2, \ldots, x_n$ probably in some different order. So, taking products of these two sets of elements, we can certainly deduce

$$a \star x_1 \star a \star x_2 \star \ldots \star a \star x_n = x_1 \star x_2 \star \ldots \star x_n$$

or

$$a^n \star x_1 \star x_2 \star \ldots \star x_n = x_1 \star x_2 \star \ldots \star x_n.$$

Multiplying on each side by the inverse of $x_1 \star x_2 \star \ldots \star x_n$ then gives

$$a^n = e. \tag{A1.1}$$

We shall later apply (A1.1) to finite fields.

⟩1.2 Rings and integral domains

These are sets with two distinct ways of combining elements. A set R, with two operations $\oplus$ and $\circledast$, is a *ring* if the following conditions are satisfied

(i) if a and b are in R, then $a \oplus b$ is also in R;

(ii) if a, b and c are in R, then $a \oplus (b \oplus c) = (a \oplus b) \oplus c$;

(iii) there is a special element $0 \in R$, called the *additive identity* of R, such that
$$a \oplus 0 = a \text{ for any } a \in R;$$

(iv) for each $a \in R$ there is a corresponding element $-a \in R$, the *additive inverse* of a, such that
$$a \oplus -a = 0;$$

(v) if a and b are in R then $a \oplus b = b \oplus a$;

(vi) if a and b are in R, then $a \circledast b$ is also in R;

(vii) if a, b and c are in R, then $a \circledast (b \circledast c) = (a \circledast b) \circledast c$;

(viii) the operation $\circledast$ is *distributive* with respect to $\oplus$: if a, b, c are any three elements in R

$$a \circledast (b \oplus c) = (a \circledast b) \oplus (a \circledast c) \text{ and } (a \oplus b) \circledast c = (a \circledast c) \oplus (b \circledast c).$$

The archetypal example of a ring is the set $\mathbb{Z}$, with $\oplus$ and $\circledast$ being the usual operations of addition and multiplication of integers. In a general ring the operations may not be ordinary addition and multiplication, and the elements may not be ordinary numbers. However, $\oplus$ and $\circledast$ still play roles analogous to those of $+$ and $*$ in the set of integers and the ring operations are easier to work with if we think of them as 'addition' and 'multiplication' and write them as $+$ and $*$ respectively. In practise confusion is unlikely to result if we are always careful to remind ourselves of the context. For each m an example of a finite ring with m elements is the set of integers $0, 1, \ldots, m - 1$ with both addition and multiplication defined modulo m as in section 1.1 above. It is denoted by $\mathbb{Z}_m$. Notice that the first five ring properties say that a ring is always an Abelian group with respect to addition; but there is no requirement for multiplication to satisfy a commutative property, for there to be a multiplicative identity nor, if there is one, for each element to have a multiplicative inverse. The ring $\mathbb{Z}$ for instance has a multiplicative identity, 1, and multiplication is commutative, but no integer other than 1 and -1 has a multiplicative inverse in $\mathbb{Z}$. The ring of all even integers, with ordinary addition and multiplication, is an example where there is no multiplicative identity. Multiplication will be commutative in all the rings we use and most will also have a multiplicative identity.

If we are working in a ring that has a multiplicative identity 1 and if for an element u we can find a u' in the ring such that

$$u * u' = u' * u = 1 \qquad (A1.2)$$

then we say that u is a *unit* or *invertible* element. In other words u is a unit when it has an inverse in the ring. The identity 1 will always be a unit since $1 * 1 = 1$ and in the ring $\mathbb{Z}_2[x]$ it is the only unit (section 1.5 in Chapter 1 showed that there are $p - 1$ units in $\mathbb{Z}_p[x]$). The ring $\mathbb{Q}[x]$ and the ring of integers in $\mathbb{Q}(\sqrt{2})$ (see Chapter 4) have infinitely many units. If u_1 and u_2 are units in some ring, with $u_1 u_1' = u_1' u_1 = 1$ and $u_2 u_2' = u_2' u_2 = 1$, then $u_1 u_2$ is also a unit because $u_1 u_2 (u_2' u_1') = u_1 (u_2 u_2') u_1' = u_1 * 1 * u_1' = 1$. The set of units is therefore closed under multiplication and multiplication of units is associative since it is associative throughout a ring. We have already pointed out that the multiplicative identity of the ring is a unit; and if u is a unit satisfying (A1.2) then u' is the inverse of u and is also a unit (as *its* inverse is u). This means that the units of a ring always form a group and it can either be finite or infinite.

In a ring R suppose we have an equation of the form

$$a + x = a + y \qquad (A1.3)$$

where a, x, y are elements of R. It is easy to see from the properties of addition that this equation implies $x = y$ no matter what value a has. For adding $-a$ to both sides of (A1.3) gives $0 + x = 0 + y$ and so $x = y$. However, it is possible to have corresponding multiplicative equations

$$a * x = a * y \qquad (A1.4)$$

with $x \neq y$. Certainly this is true when $a = 0$, because $0 = 0 + 0$ and so $0 * x = (0 + 0) * x = 0 * x + 0 * x$. Adding $-(0 * x)$ to each side of this last equation gives $0 = 0 * x$ for every x; as also $0 = 0 * y$ for every

y whether or not $y = x$. Similarly $x * 0 = y * 0 = 0$ for every x and y. More surprisingly perhaps, equation (A1.4) can also hold sometimes when $a \neq 0$ and $x \neq y$. We can see an instance of this in the ring $\mathbb{Z}_6$ which consists of the numbers $0, \ldots, 5$ with addition and multiplication defined modulo 6. Here $2*4 = 2 * 1$, since in $\mathbb{Z}$ we have $2*4 \equiv 2 * 1$ (mod 6). Some rings such as $\mathbb{Z}$ have the property that

$$if\ a \neq 0\ and\ a * x = a * y\ then\ x = y. \qquad (A1.5)$$

and are then said to satisfy a multiplicative *cancellation law*. If multiplication in the ring is not commutative we could call (A1.5) a left cancellation law and distinguish it from the right cancellation law 'if $a \neq 0$ and $x * a = y * a$ then $x = y$'. These two versions are the same for all rings with commutative multiplication and so we shall not separate them. A ring in which multiplication is commutative, which has a multiplicative identity and in which the cancellation law (A1.5) holds is an *integral domain*. The ring of integers $\mathbb{Z}$, the ring of rational numbers $\mathbb{Q}$, and the finite ring $\mathbb{Z}_p$ for a prime p, are all integral domains. In $\mathbb{Z}_p$ the multiplicative cancellation law holds because if $a \in \mathbb{Z}_p$ is not zero and $a * x = a * y$ in $\mathbb{Z}_p$ then $ax \equiv ay$ (mod p) in $\mathbb{Z}$ with $p \nmid a$. From the properties of congruences, see Jackson(1987), there will be a z such that $za \equiv 1$ (mod p) and then $x \equiv zax \equiv zay \equiv y$ (mod p), meaning $x = y$ in $\mathbb{Z}_p$.

Another property that is equivalent to the cancellation law is

$$if\ a * b = 0\ then\ either\ a = 0\ or\ b = 0. \qquad (A1.6)$$

In order to see that (A1.5) and (A1.6) are equivalent suppose first that the cancellation law holds, that $a * b$ is a product which is zero, and that $a \neq 0$. We must then show that $b = 0$. As in the remarks after (A1.4) we have $a * 0 = 0$, so that $a * b = a * 0$. The cancellation law applied to this last equation implies $b = 0$ which is the conclusion of (A1.6). On the other hand, suppose that (A1.6) always holds and that $a * x = a * y$ for some $a \neq 0$. Then $a * x - a * y = 0$, or $a * (x - y) = 0$. Since $a \neq 0$, property (A1.6) applied to this equation gives $x - y = 0$ or $x = y$, and that is the statement of (A1.5). So an integral domain can alternatively be regarded as a commutative ring with identity in which (A1.6) holds. Notice that (A1.6) can easily be extended by induction to give

$$if\ a_1 * a_2 * \ldots * a_n = 0\ \ then\ a_1 = 0\ or\ a_2 = 0\ or\ \ldots\ or\ a_n = 0. \quad (A1.7)$$

⟩1.3 Divisibility in integral domains

If D is an integral domain and a, b are elements in D then b divides a if there is an element $c \in D$ with

$$a = bc. \tag{A1.8}$$

A unit u of D divides every element, because if u' is the inverse of u then $a = u(u'a)$. This also shows that the element a is divisible by any unit multiple of itself, and unit multiples of an element are called *associates* of it. For example, in the domain $\mathbb{Z}$, -7 is an associate of 7 and in the domain of quadratic integers of $\mathbb{Q}(\sqrt{-1})$ the units are $1, -1, i = \sqrt{-1}$ and $-i$, and the associates of 7 there are $7, -7, 7i, -7i$. In any domain a *common divisor* of two elements is an element that divides both of them; and if there is a common divisor that is divisible by each other common divisor it is called a *greatest common divisor*, or GCD of the elements. Thus 1 is a common divisor of any two elements of a domain, and in $\mathbb{Z}$ the pair $0, 0$ does not have a GCD because every non-zero element is a common divisor in that case. If g is a GCD of two elements e and f in an integral domain, then every other GCD of e and f must be a unit multiple of g. In order to see this, suppose that g' is another GCD of e and f. Then, by definition, g' will be a multiple, say ug, of the common divisor g; and the GCD g will be a multiple of the common divisor g', say $g = u'g'$. So $g' = ug = u(u'g')$, whence $uu' = 1$ meaning that u is a unit and g' is an associate of g. Conversely, any associate of a GCD will also be a GCD. If say u_1 is a unit with inverse u_1', then $u_1 g$ divides into each of e and f since $g = u_1'(u_1 g)$ does; and $u_1 g$ is divisible by all other common divisors since they all divide g. In $\mathbb{Z}[\sqrt{-1}]$ for instance, a GCD of $14 + 6i$ and 16 is $2 + 2i$ (see section 3.4), and the other greatest common divisors of $14 + 6i$ and 16 are then $-1 * (2 + 2i) = -2 - 2i$, $i * (2 + 2i) = -2 + 2i$ and $-i * (2 + 2i) = 2 - 2i$.

Just as in $\mathbb{Z}$ a non-zero and non-unit element of a general integral domain is *irreducible* if it is only divisible by units and associates of itself; otherwise it is *composite*. Equivalently, a non-zero, non-unit element is irreducible if the only way of writing it as a product is for one of the factors to be a unit; and a composite element can be written as a product of two terms neither of which is a unit. In $\mathbb{Z}$ both 13 and -13 are irreducible and among the quadratic integers in $\mathbb{Q}(\sqrt{-1})$ the elements $7 + 2i, -7 - 2i, -2 + 7i$ and $2 - 7i$ are all irreducible (see chapter 3). In a domain D there may be some elements p that satisfy another closely related property: that if p divides a product of two terms then it necessarily divides one of the factors. In other words, if a and b are in D then

$$p|ab \; implies \; p|a \; or \; p|b. \tag{A1.9}$$

An element p satisfying (A1.9) is *prime* in the domain and in $\mathbb{Z}$ both the positive and negative irreducible elements have this prime property (see e.g. Jackson(1987)). Notice that if p satisfies (A1.9) then for any $n \geqslant 2$ it satisfies

$$p|a_1 a_2 \ldots a_n \; implies \; p|a_1 \; or \; p|a_2 \; or \; \ldots \; or \; p|a_n; \tag{A1.10}$$

because if (A1.9) holds and p divides $a_1 * a_2 \ldots a_n$ then it must divide a_1 or $a_2 * \ldots a_n$, and so divide a_1 or a_2 or $a_3 \ldots a_n$, and, by continued application of (A1.9), either $p|a_1$ or $p|a_2$ or $\ldots$ or $p|a_n$.

In any domain every prime element is irreducible. To see this let us suppose that p is prime in a domain D and that it can be written as a product bc where b and c are in D. Then, from the prime property, $p = bc$ implies that either p divides b or p divides c. If p were to divide b then b would be pb_1 for some $b_1 \in D$, giving $p = bc = pb_1 c$. That would mean $p * 1 = p * b_1 c$ and the cancellation property would imply $1 = b_1 c$ whence c would be a unit. The alternative choice $p|c$ leads to b being a unit. Therefore the only way that p can be written as a product bc is for one of the factors to be a unit (and the other to be an associate of p). So p is irreducible.

〉1.4 Euclidean domains and factorization into irreducibles

In $\mathbb{Z}$, as we have remarked, irreducible elements are also prime; and one of the consequences is that every rational integer has a unique factorization into irreducible integers. Corresponding statements are true in several other integral domains, but there are some domains such as the ring of integers in $\mathbb{Q}(\sqrt{-5})$ where there are irreducible elements that are not prime (see Chapter 4). In $\mathbb{Z}$ the proof that irreducible elements also have the prime property (A1.9) turns out to depend ultimately on the possibility of performing division with a remainder that is smaller than the divisor. Many domains have no similar ordering of their elements and no concept of being 'smaller', but it is possible to slightly generalize the division with remainder property in such a way that it applies to some other integral domains and still remains useful. Instead of looking for an ordering of the elements in a domain we shall look for a way of giving each element an integer size. Specifying an integer size for each element is the same as specifying a function from the domain to the integers, and a function with the following three properties encapsulates the division with remainder property of $\mathbb{Z}$ while being more widely applicable. A *Euclidean norm* or *Euclidean valuation* on an integral domain D is a function v from the non-zero elements of D to $\mathbb{Z}$ and which satisfies

(i) $v(a) \geqslant 0$ for every non-zero a in D;

(ii) $b|a$ implies $v(b) \leqslant v(a)$ if a and b are both non-zero;

(iii) if a, b are in D and $b \neq 0$ then there are elements q and r in D such that

$$a = qb + r \text{ where either } r = 0 \text{ or } v(r) < v(b).$$

An integral domain that has a Euclidean norm is called a *Euclidean domain*. In property (ii) we can only attempt division by non-zero elements so we are not interested in whether v is defined at 0 or not. The domain $\mathbb{Z}$ is a Euclidean domain since a suitable Euclidean norm is given by putting $v(a) = |a|$ for each non-zero integer a. Chapter 1 shows that the domains $\mathbb{Q}[x]$ and $\mathbb{Z}_p[x]$ are also Euclidean domains with the degree function as Euclidean norm; and chapter 4 shows that the domain of quadratic integers in $\mathbb{Q}(\sqrt{d})$ is a Euclidean domain when $d = -11$, $-7, -3, -2, -1, 2, 3, 5$ and the computer will help with investigations into the domains with $|d| \leqslant 5$.

In a Euclidean domain the multiplicative identity 1 always has the smallest possible norm. This is because 1 divides each element a and so $v(1) \leqslant v(a)$, where v is the Euclidean norm. Each unit shares this norm, since if u is a unit with inverse u' then of course $v(1) \leqslant v(u)$ and also $u * u' = 1$ implies $v(u) \leqslant v(1)$, giving $v(u) = v(1)$. Moreover if an element e has this same smallest norm then it must be a unit: if e did not divide exactly into 1 we could write $1 = qe + r$ with $v(r) < v(e) = v(1)$ contrary to the fact that there is no norm smaller than $v(1)$. For instance in $\mathbb{Q}[x]$ or $\mathbb{Z}_p[x]$ the units have degree zero, the smallest possible degree there.

Now suppose that a divides b in the domain: say $b = at$. Then $v(a) \leqslant v(b)$. If b also divides a, so that a and b are associates, then $v(b) \leqslant v(a)$ whence $v(a) = v(b)$. If a divides b but b does not divide a, we can write $a = qb + r$ for a non-zero r with $v(r) < v(b)$. Then $r = a - qb = a(1 - qt)$ and this means that $v(a) \leqslant v(r) < v(b)$. In other words *associates have the same norm, but*

divisors that are not associates have strictly smaller norms.

Using this we can deduce for example that in a Euclidean domain any element with the second smallest norm must be irreducible. For if such an element, b say, had a divisor that was not one of its own associates, then that divisor, being of smaller norm than b, would have to have norm $v(1)$ and so be a unit. So b is not a unit but only has units and associates as divisors and is therefore irreducible. (Compare the proof in Example 1.5 that linear polynomials in $\mathbb{Q}[x]$ and $\mathbb{Z}_p[x]$ are irreducible.)

We can also show that in an arbitrary Euclidean domain D each non-zero element is either a unit, irreducible or a finite product of irreducibles. We use induction on the norms of the elements of D. As we have seen, those elements of smallest norm are units. Suppose then that we have reached the stage of having proved that for some rational integer $n \geqslant v(1)$ every non-zero element with norm not exceeding n is either a unit, irreducible or a product of irreducibles. Consider the next possible norm $n + 1$. If there are no composite elements with norm $n + 1$ the inductive hypothesis is satisfied with n replaced by $n + 1$. Otherwise let a be a composite element in D of norm $n + 1$. Then, by definition, a will be the product of two non-zero elements b and c in D, neither of which is a unit. So from the previous paragraph $v(b) < v(a)$ and $v(c) < v(a)$. That is, $v(b) \leqslant n$ and $v(c) \leqslant n$, whence, by assumption, b and c are each either irreducible or can be written as products of irreducibles. Writing $b = b_1 \ldots b_r$ and $c = c_1 \ldots c_s$, (where $b_1, \ldots, b_r, c_1, \ldots, c_s$ are irreducible and $r = 1$ or $s = 1$ if b or c is irreducible) we have $a = bc = b_1 \ldots b_r c_1 \ldots c_s$. This gives the inductive step from norms not exceeding n to norms not exceeding $n + 1$ and the result is therefore true without restriction.

Note that in the argument above we only used the division with remainder property in two places: to show that an element with the smallest possible norm must be a unit and also that a divisor of an element b, which is not an associate of b, must have a strictly smaller norm than b. So if we could establish those properties in another way, and if we were working in an integral domain with a norm function v that satisfied the requirements $v(a) \geqslant 0$ for every non-zero a and $b|a$ implies $v(b) \leqslant v(a)$ if a and b are both non-zero, then we could still deduce that each non-zero element is either a unit, irreducible or a finite product of irreducibles. This is what we can do in domains of quadratic integers (see Chapter 4).

They are not all Euclidean domains, but they all have norms that are multiplicative and, in some situations, that is just as good.

In any Euclidean domain we can also use division with remainder to show that

> *any two non-zero elements a, b have a greatest common divisor*
>
> *and every GCD of a and b can be written as*
>
> *a linear combination of them*

If a, b are in a domain D with Euclidean norm v, we consider the set $S = \{ax + by | x, y \in D\}$ of all linear combinations of a and b, with coefficients in D. The element $0 = ab + b(-a)$ is in S, but we shall be more interested in non-zero elements that can be written as linear combinations of a and b. There certainly are some, because $a = a*1 + b*0$ and $b = a*0 + b*1$ are both in S. Since all norms are non-negative integers, there will be one or more of these combinations that have least possible norm out of all the non-zero elements of S. Let $d = ax_0 + by_0$ be such an element in S of least norm: it will turn out that d is a GCD of a, b. First we use division with remainder to write

$$a = qd + r \text{ where either } r = 0 \text{ or } v(r) < v(d). \qquad (A1.11)$$

Now $r = a - qd = a(1 - qx_0) + (-q)y_0$ is a linear combination in S, so if r were non-zero we would have $v(r) \geqslant v(d)$ as d is one of the elements of least norm in S. This would contradict (A1.11). Thus r is necessarily zero, meaning that d divides a. By similarly dividing d into b with a remainder that is either zero or has smaller norm than d does, we see that $d \mid b$. So d is a common divisor of a, b; and any other common divisor must divide the linear combination $ax_0 + by_0 = d$. In other words d is a GCD of a, b. If now d_1 is any other GCD it will be a unit multiple of d (as shown at the beginning of this section), so that $d_1 = ud = a(ux_0) + b(uy_0)$ will also be of the required form. In practise we would find a GCD of two elements in a specific domain by using the method of repeated division with remainder known as Euclid's algorithm (see sections 1.4 and 3.4).

⟩1.5 Unique factorization in Euclidean domains

These facts about greatest common divisors enable us to show that the irreducible elements in a Euclidean domain are also prime. In order to see this suppose that p is an irreducible element of a Euclidean domain D and that p divides a product ab in D. We must show that p divides one of the factors and we do this by supposing that p does not divide a and prove that it must then divide the other factor b. Now every common divisor of p and a has to in particular be a divisor of p and so be either a unit or an associate of p, as those are the only types of divisors of irreducibles. Any associate of p is of the form vp for a unit v, and if vp were a divisor of a then p would also divide a. So, under the assumption that p does not divide a, we see that units are the only common divisors of p and a. If u is a greatest common divisor of p, a, it must therefore be a unit. The previous paragraph now implies that there will be elements x, y in D such that

$$ax + py = u.$$

Multiplying both sides by the inverse u' of u gives

$$au'x + pu'y = 1$$

and then

$$abu'x + pbu'y = b.$$

Since p divides ab by our initial assumption, it divides all of the left side of this last equality and so must divide b. This means that p has the prime property (A1.9) and this is true of every irreducible element in a Euclidean domain.

We can now show that in a Euclidean domain two expressions for a composite element as a product of primes (or equivalently, of irreducibles) must have the same number of factors, and each prime occurring in either one of the expressions must be a unit multiple of some prime in the other. Again we prove this by induction. So, as before, let D be a Euclidean domain with Euclidean norm v. We know that those integers with norm $v(1)$ are units. Suppose we have proved that every non-zero element with norm less than some $n > v(1)$ is either a unit, a prime or has an essentially unique expression as a product of primes and that a is an element in D of norm n which can be represented as $p_1 p_2 \ldots p_r$ and also as $q_1 q_2 \ldots q_s$ for primes $p_1, \ldots, p_r, q_1, \ldots, q_s$. From $p_1 p_2 \ldots p_r = q_1 q_2 \ldots q_s$, we see, using (A1.10), that p_1 must divide one of $q_1, \ldots, q_s$ and by relabeling $q_1, \ldots, q_s$ if necessary we may put $q_1 = up_1$; and the only way that can hold, with p_1 and q_1 irreducible, is for u to be a unit. Then $p_2 \ldots p_r = (uq_2) \ldots q_s$ and this, being $\frac{a}{p_1}$ and so a divisor but not an associate of a, is an integer of smaller norm than n. The assumption about the prime factors of integers with norm less than n now implies that $r = s$ and uq_2 (and so q_2), ..., q_s are each unit multiples of $p_2, ..., p_r$, though perhaps in a different order. This provides the inductive step from factorizations of elements with norm less than n to factorizations of those with norm n. So all composite elements have essentially unique factorizations into primes.

⟩1.6 Integral domains and fields

A field is a further type of ring in which the properties of multiplication mirror all those of addition. That is, a set F, with two operations $+$ and $*$, is a *field* if the following conditions are satisfied

(i) if a and b are in F, then $a + b$ is also in F;

(ii) if a, b and c are in F, then $a + (b + c) = (a + b) + c$;

(iii) there is a special element $0 \in F$ such that

$$a + 0 = a \text{ for any } a \in F;$$

(iv) for each $a \in F$ there is a corresponding element $-a \in F$ such that

$$a + (-a) = 0;$$

(v) if a and b are in F then $a + b = b + a$;

(vi) if a and b are in F, then $a * b$ is also in F;

(vii) if a, b and c are in F, then $a * (b * c) = (a * b) * c$;

(viii) there is an element $1 \in F$ (the *multiplicative identity* of F) such that $1 \neq 0$ and

$$a * 1 = a \text{ for any } a \in F;$$

(ix) for each non-zero element $a \in F$ there is a corresponding element $a^{-1} \in F$ (the *multiplicative inverse* of a) such that

$$a * (a^{-1}) = 1;$$

(x) if a and b are in F then $a * b = b * a$;

(xi) the operation $*$ is distributive with respect to $+$: if a, b, c are any three elements in F

$$a * (b + c) = (a * b) + (a * c) \text{ and } (a + b) * c = (a * c) + (b * c).$$

These properties mean that a field is an Abelian group with respect to addition and also the non-zero field elements form an Abelian group with respect to multiplication. The set of rational numbers, with their usual addition and multiplication, form a field; as do similarly the sets of real and complex numbers with their addition and multiplication. For each prime p the integers $0, 1, \ldots, p - 1$ with the operations of addition modulo p and multiplication modulo p, as in section 1.1 above, form a finite field $\mathbb{Z}_p$ (see also the end of section 1.2).

Every field is an integral domain; for suppose we have an equation $a * x = a * y$ with $a \neq 0$ as in (A1.5). Then a will have a multiplicative inverse a^{-1} and we can multiply both sides of the equation by it to obtain $(a^{-1} * a) * x = (a^{-1} * a) * y$, and so $1 * x = x = 1 * y = y$ as required. Conversely, if an integral domain is finite then it must be a field. In order to see this suppose that D is a finite integral domain. Then D is a ring in which multiplication is commutative and where there is a multiplicative identity; so it satisfies all the field axioms except perhaps for the existence of inverses of non-zero elements. So let the elements of D be $a_1, \ldots, a_n$ and consider all the multiples $a_i * a_1, a_i * a_2, \ldots, a_i * a_n$, of one of the non-zero elements a_i. They are all distinct because, by the cancellation law in D, $a_i * a_j = a_i * a_k$ could only happen if $a_j = a_k$, since $a_i \neq 0$. The n multiples must therefore be all the n elements of D. In particular there must be one multiple, say $a_i * a_l$, that is equal to 1. This a_l is then the required inverse of a_i. Each non-zero element similarly has a multiplicative inverse and D is thus a field. There are (necessarily infinite) integral domains that are not fields: the ring $\mathbb{Z}$ is one such. There are no prime or composite elements in a field because every non-zero element is invertible. So divisibility questions are less interesting in fields than in other integral domains.

$\rangle$1.7 Finite fields

In a finite field with q elements we have two finite groups to work with: the additive group of all q elements and the multiplicative group of the $q - 1$ non-zero elements. The identity of the first group is 0 and for any element x the sum $x + x + \ldots + x$ with q summands can be conveniently written as qx. So, when translated into additive notation, (A1.1) becomes

$$qx = 0 \qquad \text{(A1.12)}$$

for every x in the field. In the multiplicative case (A1.1) becomes

$$x^{q-1} = 1 \qquad \text{(A1.13)}$$

for every non-zero x in the field. Multiplying both sides of (A1.13) by x we get

$$x^q = x \qquad \text{(A1.14)}$$

and this is true not only for every non-zero x but also for $x = 0$. [Just after (A1.4) we saw that a product involving 0 is equal to 0 in any ring (or field). In particular $0 * 0 = 0$, and $0 * 0 * \ldots * 0 = 0$.] In the case of the field $\mathbb{Z}_p$ for a prime p, (A1.13) gives $x^{p-1} = 1$ for all non-zero x in $\mathbb{Z}_p$, or equivalently

$$x^{p-1} \equiv 1 \ (\text{mod } p) \qquad \text{(A1.15)}$$

for those integers x that are not divisible by p. Similarly, with $q = p$, (A1.14) applied to $\mathbb{Z}_p$ gives

$$x^p \equiv x \ (\text{mod } p) \qquad \text{(A1.16)}$$

for every x in $\mathbb{Z}$. The congruences (A1.15) and (A1.16) are originally due to P. Fermat (1601–1665).

Notice that some of these identities can be easily extended. The additive identity (A1.12) for instance implies that

$$nqx = 0 \qquad \text{(A1.17)}$$

for any positive integer n. Also, by raising both sides of (A1.14) to the power q we get $x^{q^2} = (x^q)^q = x^q = x$; and repeatedly raising both sides to the power q leads to

$$x^{q^n} = x \qquad \text{(A1.18)}$$

which again holds for any positive n and any x in the field. In particular (A1.16) similarly leads to

$$x^{p^n} \equiv x \ (\text{mod } p) \qquad \text{(A1.19)}$$

for every prime p and $x \in \mathbb{Z}$.

The identity (A1.12) places great constraints on the possible structure of a finite field. First of all, in a field F with q elements, (A1.12) holds when x is the multiplicative identity 1 so that

$$q1 = 1 + 1 + \ldots + 1 = 0. \qquad \text{(A1.20)}$$

Now for any positive integers m and n, the distributive property gives

$$1 * n1 = 1 * (1 + 1 + \ldots + 1) = 1 * 1 + 1 * 1 + \ldots$$

$$+ 1 * 1 = 1 + 1 + \ldots + 1 = n1$$

and then repeated use of distributivity gives

$$(m1) * (n1) = (1 + \ldots + 1) * (n1) = (1 * n1) + (1 * n1) + \ldots + (1 * n1)$$

$$= n1 + \ldots + n1 = mn1.$$

So, if $q = p_1 p_2 \ldots p_r$ where $p_1, p_2, \ldots, p_r$ are primes (not necessarily all different), (A1.20) implies

$$q1 = (p_1 1) * (p_2 1) * \ldots * (p_r 1) = 0.$$

Then (A1.7), which is always true in a field, means that one of these factors must be zero. Dropping the subscripts, we have shown that there is some prime p for which

$$p1 = 1 + \ldots + 1 = 0. \qquad (A1.21)$$

We could not have $k1 = 0$ for any integer k which was not a multiple of p, since for such a k there would be integers s and t with $sp + tk = 1 \in \mathbb{Z}$ and then $s(p1) + t(k1) = 0 \in F$ and also $s(p1) + t(k1) = (sp + tk)1 = 1 \in F$, contradicting the fact that 1 and 0 are different in a field. So the set

$$F_0 = \{0, 1, 1 + 1, \ldots, (p - 1)1\} \qquad (A1.22)$$

consists of p distinct elements (since $r_1 1 = r_2 1$ with $0 \leqslant r_1 < r_2 < p$ would give $(r_2 - r_1)1 = 0$ with $0 < r_2 - r_1 < p$). It is also closed with respect to the addition and multiplication in F; because for $m1$ and $n1$ in F_0, we have $m1 + n1 = (m + n)1$ and if $m + n \equiv r$ (mod p) then $(m + n)1 = r1 \in F_0$. Similarly $(m1) * (n1) = mn1 \in F_0$. The additive inverse of an element $m1$ in F_0 again lies in F_0 because $-m1 = (p - m)1 \in F_0$. Finally if $m1$ is a non-zero element of F_0, so $0 < m < p$, then there will be an integer u with $0 < u < p$ such that $mu \equiv 1 \in \mathbb{Z}$ (mod p) meaning that $(m1) * (u1) = 1 \in F_0$. The other field axioms (the associativity and commutativity of addition and multiplication and the distributivity of multiplication over addition) are true in F_0 because they hold throughout F. So F_0 is a subfield of F.

Moreover, it is a subfield that is isomorphic to Z_p. The easiest way to see this is to imagine labelling the field elements 0 and 1 in the definition of F_0 with the integers 0 and 1, the field element $1 + 1$ with 2, $\ldots$, and the field element $(p - 1)1$ with the integer $p - 1$. This associates the set of integers $0, 1, 2, \ldots, p - 1$ with the elements of F_0. Then the field elements $m1$ and $n1$ are associated with the integers m and n and, because $m1 + n1 = r1$ where $0 \leqslant r < p$ and $r \equiv m + n \pmod{p}$, $m1 + n1$ is associated with r where r is the result of adding m and n modulo p. That is, we can add elements of F_0 by just adding their labels in $\mathbb{Z}_p$ and translating the result back into F_0. Similarly we can multiply elements of F_0 by multiplying their labels modulo p and translating the result into F_0. So we can do all arithmetic in F_0 by working entirely in $\mathbb{Z}_p$ and just changing names as appropriate. In other words we can regard the elements of F_0 as *being* the elements of $\mathbb{Z}_p$ with different names. So we have shown that, for some prime p, the finite field F is an extension field of $\mathbb{Z}_p$.

In the rest of this section we shall show that the number of elements in a finite field has to be a power of a prime. So let F be a finite field and, as above, F_0 its subfield which is isomorphic to $\mathbb{Z}_p$ for some prime p. If every element in F can be expressed as a sum of the form $1 + 1 + \ldots + 1$ then $F = F_0$ and F contains p elements.

If not, there must be an element a_1 of F which is not in F_0. Consider all the linear combinations of 1 and a_1 with coefficients in $\mathbb{Z}_p$:

$$F_1 = \{r1 + sa_1 | 0 \leqslant r < p, \ 0 \leqslant s < p\}.$$

Clearly $F_0 \subset F_1 \subseteq F$ and we can show that the combinations in F_1 are all distinct. Certainly we could not have $r1 + sa_1 = r'1 + sa_1$ with $r \neq r'$, for that would imply $r1 = r'1$, which could not happen as all the elements in F_0 are distinct. Neither could we have $r1 + sa_1 = r'1 + s'a_1$ with $s \neq s'$ for that would imply $(r - r')1 = (s' - s)a_1 = (s' - s)1a_1$. The inverse of the non-zero element $(s' - s)1 \in F_0$ is again in F_0 so multiplying through by that inverse would give the contradiction $a_1 \in F_0$. If $F = F_1$, the number of elements in F is then the number of elements in F_1 which is p^2.

Otherwise $F \neq F_1$ and we choose any element a_2 not in F_1 and consider the set

$$F_2 = \{r1 + sa_1 + ta_2 | 0 \leqslant r < p, \ 0 \leqslant s < p, \ 0 \leqslant t < p\}.$$

Then $F_1 \subset F_2 \subseteq F$ and we can show similarly that the combinations in F_2 are all distinct. We could not have $r1 + sa_1 + ta_2 = r'1 + s'a_1 + ta_2$

with $(r, s) \neq (r', s')$ as all the combinations in F_1 are distinct. Neither could we have $r1 + sa_1 + ta_2 = r'1 + s'a_1 + t'a_2$ with $t \neq t'$, for then $(t - t')1a_2$ would be an element of F_1 which would lead, as before, to the contradiction $a_2 \in F_1$. So there are p^3 elements in F_2 and if $F = F_2$ they are the p^3 elements of F.

If $F \neq F_2$ we continue by choosing any element a_3 not in F_2 and form the set F_3 of linear combinations of $1, a_1, a_2, a_3$. Suppose that we have in this way formed a sequence of elements $1, a_1, \ldots, a_i$ in F and the corresponding sequence of sets $F_0, F_1, \ldots, F_i$, where each set F_j consists of the linear combinations of $1, a_1, \ldots, a_j$ with coefficients from $\mathbb{Z}_p$, and that distinct linear combinations have distinct values. Then if $F \neq F_i$ we choose an element $a_{i+1} \notin F_i$ and form

$$F_{i+1} = \{r1 + sa_1 + \ldots + va_{i+1} | 0 \leqslant r < p, 0 \leqslant s < p, \ldots, 0 \leqslant v < p\}.$$

It is again straightforward to show that different linear combinations in F_{i+1} must have different values. If for instance we had $r1 + sa_1 + \ldots + va_{i+1} = r'1 + s'a_1 + \ldots + va_{i+1}$ then that would imply that two different linear combinations in F_i were equal. If we had $r1 + sa_1 + \ldots + va_{i+1} = r'1 + s'a_1 + \ldots + v'a_{i+1}$ for $v \neq v'$ then that would imply that $(v - v')1a_{i+1}$ was an element of F_i, and, as before, that a_{i+1} was in F_i. So in the sequence $F_0, F_1, \ldots$ each set F_j contains p^j elements, as there are p^j different linear combinations in F_j, and the sequence may be continued as long as its last member is not equal to F. Since F has only a finite number of elements, the sequence cannot grow indefinitely and therefore there will be a k with $F = F_k$. Then the number of elements in F is p^k.

A polynomial is called primitive if it has integer coefficients that have GCD 1 (see Chapter 1, section 1.7). For example $2x^5 - 3x^2 + 7$ and $4x^3 + x^2 - 13$ are each primitive. Their sum is $2x^5 + 4x^3 - 2x^2 - 6$ which is not primitive since every coefficient is even. However their product is $8x^8 + 2x^7 - 38x^5 - 3x^4 + 28x^3 + 46x^2 - 91$ which is primitive. Although sums of primitive polynomials are sometimes primitive and sometimes not, *the product of two primitive polynomials is always primitive.* This result is due to C. F. Gauss (1777–1855) and is often known as *Gauss' Lemma.* Suppose then that f and g are two primitive polynomials with say

$$f(x) = a_n x^n + a_{n-1} x^{n-1} + \ldots + a_0$$

and

$$g(x) = b_m x^m + b_{m-1} x^{m-1} + \ldots + b_0.$$

Their product is

$$f(x)g(x) = \sum_{k=0}^{m+n} \sum_{i=0}^{k} a_i b_{k-i} x^k \qquad \text{(A2.1)}$$

which plainly has integer coefficients. So the only way it could fail to be primitive would be if all of its coefficients had a common divisor greater than 1; and that would entail them having a common prime divisor. We shall show that no prime can divide all the coefficients of fg. So let p be a prime. Since f and g are primitive we know that p does not divide all the coefficients of f nor all those of g, though it might divide some of them. Let s be the least subscript such that p does not divide a_s. (Although p might divide $a_0, a_1, \ldots$, it does not divide every a_i so there will certainly be a first $s \leqslant n$ with $p \nmid a_s$.) Similarly let t be the least subscript such that p does not divide b_t. Then in the product fg, the coefficent of x^{s+t} is

$$\sum_{i=0}^{s+t} a_i b_{s+t-i} = \sum_{i=0}^{s-1} a_i b_{s+t-i} + a_s b_t + \sum_{i=s+1}^{s+t} a_i b_{s+t-i}$$

$$= \sum_{i=0}^{s-1} a_i b_{s+t-i} + a_s b_t + \sum_{j=0}^{t-1} a_{s+t-j} b_j$$

and every term in the two last summations is divisible by p since it divides every a_i with $0 \leqslant i \leqslant s-1$ and every b_j with $0 \leqslant j \leqslant t-1$. The remaining term $a_s b_t$ is not divisible by p since it does not divide either a_s or b_t, and so the whole coefficient is not divisible by p. We have therefore shown that no prime can divide every coefficient of fg and that is the required result.

⟩ Appendix 3

⟩ The Möbius function and cyclotomic polynomials

The Möbius function μ is named after A F Möbius (1790–1868) and is defined for each positive integer n as follows

$$\mu(n) = \begin{cases} 1 & \text{if } n = 1 \\ (-1)^r & \text{if } n \text{ is a product of } r \text{ distinct primes.} \\ 0 & \text{otherwise} \end{cases} \quad (A3.1)$$

Thus $\mu(1) = 1$, $\mu(2) = -1$, $\mu(3) = -1$, $\mu(4) = 0$, ... The Möbius function occurs in the definitions of the cyclotomic polynomials. There is one of these for each natural number k, given by

$$f_k(x) = \prod_{d|k}(x^d - 1)^{\mu(\frac{k}{d})} = \prod_{h|k}(x^{\frac{k}{h}} - 1)^{\mu(h)}. \quad (A3.2)$$

In (A3.2) the variables d and h range over all the positive divisors of k, and on putting $d = \frac{k}{h}$ it is apparent that the two products in (A3.2) are the same. It can be shown that $f_k(x)$ is always a polynomial with integer coefficients and that each f_k is irreducible in $\mathbb{Q}[x]$ or, equivalently, in $\mathbb{Z}[x]$. For example, from (A3.2),

$$f_1(x) = (x^1 - 1)^{\mu(1)} = x - 1$$
$$f_2(x) = (x^2 - 1)^{\mu(1)}(x^1 - 1)^{\mu(2)} = (x^2 - 1)(x - 1)^{-1} = x + 1$$

and

$$f_6(x) = (x^6 - 1)^{\mu(1)}(x^3 - 1)^{\mu(2)}(x^2 - 1)^{\mu(3)}(x^1 - 1)^{\mu(6)}$$
$$= (x^6 - 1)(x - 1)(x^3 - 1)^{-1}(x^2 - 1)^{-1}$$
$$= x^2 - x + 1.$$

Although it is hard to show that the general cyclotomic polynomial is irreducible, we can use Eisenstein's criterion to see this in the case of f_p for a prime p. From (A3.2)

$$f_p(x) = \frac{x^p - 1}{x - 1} = x^{p-1} + x^{p-2} + \ldots + x + 1 \qquad (A3.3)$$

and if we put $x = y + 1$ the polynomial becomes

$$F(y) = \frac{(y + 1)^p - 1}{y} = y^{p-1} + py^{p-2} + \ldots + p. \qquad (A3.4)$$

Since $F(y) = f_p(y + 1)$ any non-trivial factorization of f_p would immediately give a non-trivial factorization of F. For if $f_p(x) = g(x)h(x)$ then $F(y) = g(y+1)h(y+1)$. Similarly any factorization of F would produce a factorization of f_p. So F and f_p are both irreducible or both composite. But (A3.4) shows that, apart from its leading coefficient of 1, the coefficients of F are binomial coefficients that are divisible by p and its constant term is not divisible by p^2. So Eisenstein's criterion implies that F, and therefore also f_p, is irreducible. Note that the degree of f_p is $p - 1 = \varphi(p)$. For every k the degree of f_k turns out to be $\varphi(k)$.

⟩ Appendix 4

⟩ Rouché's theorem

Rouché's theorem (see e.g. Ahlfors(1953)) is a result in complex analysis that can be useful in testing integer polynomials for irreducibility. It says that if g and h belong to a wide class of real or complex valued functions (that includes polynomials) and if R is a positive number such that

$$|g(x)| < |h(x)| \text{ for all } x \text{ with } |x| = R \qquad \text{(A4.1)}$$

then the function $g + h$ has the same number of roots satisfying $|x| < R$ as the function h does.

In order to see how this might help us to decide that a particular polynomial is irreducible we shall suppose that $f(x) = a_n x^n + a_{n-1} x^{n-1} + \ldots + a_0$ is a polynomial with integer coefficients and let $h(x) = a_{n-1} x^{n-1}$ and $g(x) = a_n x^n + a_{n-2} x^{n-2} + \ldots + a_0$. Then $g + h = f$ and the $n - 1$ roots of h (corresponding to its $n - 1$ linear factors) are all zero. So if we find a positive number R with

$$\left| a_n x^n + a_{n-2} x^{n-2} + \ldots + a_0 \right| < \left| a_{n-1} x^{n-1} \right| \quad \text{whenever } |x| = R \quad \text{(A4.2)}$$

then from Rouché's result we can deduce that f has $n - 1$ roots satisfying $|x| < R$, since h does. For example the polynomial $x^5 + 7x^4 - 3x^2 + x + 1$ has 4 roots satisfying $|x| < 1$. This is because we can take $h(x) = 7x^4$, $g(x) = x^5 - 3x^2 + x + 1$ and then with $|x| = 1$ we have $|h(x)| = |7x^4| = 7$ and $|g(x)| = |x^5 - 3x^2 + x + 1| \leqslant |x^5| + |-3x^2| + |x| + 1 = 6 < |h(x)|$. Similarly it is easy to see that the polynomial $2x^4 + 5x^3 + x + 1$ has 3 roots satisfying $|x| < 0.9$; because when $|x| = 0.9$ we have $|5x^3| = 3.645$ and $|2x^4 + x + 1| \leqslant |2x^4| + |x| + 1 = 3.2122$.

These observations do not by themselves imply irreducibility; but suppose that the polynomial f has integer coefficients, with leading coefficient 1 and non-zero constant term, and that we are able to satisfy (A4.2) with $R = 1$. Then f has $n - 1$ roots with modulus less than 1. Also if $\alpha_1, \alpha_2, \ldots, \alpha_n$ are all the roots of f then

$$f(x) = (x - \alpha_1) \ldots (x - \alpha_n) \tag{A4.3}$$

since each linear factor is a divisor of f and both sides have leading coefficient 1. If now f were composite, it could be written as a product of two polynomials with integer coefficients, say $x^r + \ldots + b_0$ and $x^s + \ldots + c_0$ where $r + s = n$. So we would have

$$(x^r + \ldots + b_0)(x^s + \ldots + c_0) = (x - \alpha_1) \ldots (x - \alpha_n). \tag{A4.4}$$

We could then relabel the roots, if necessary, so that $x^r + \ldots + b_0 = (x - \alpha_1) \ldots (x - \alpha_r)$ and $x^s + \ldots + c_0 = (x - \alpha_{r+1}) \ldots (x - \alpha_n)$, and comparing the constant terms on each side of these equations would give

$$|\alpha_1 \ldots \alpha_r| = |b_0| \geqslant 1 \quad \text{and} \quad |\alpha_{r+1} \ldots \alpha_n| = |c_0| \geqslant 1. \tag{A4.5}$$

But we know from Rouché's result that all but one of the roots have moduli less than 1. So for any r with $1 \leqslant r < n$ we must have either $|\alpha_1 \ldots \alpha_r| = |\alpha_1| \ldots |\alpha_r| < 1$ or $|\alpha_{r+1} \ldots \alpha_n| = |\alpha_{r+1}| \ldots |\alpha_n| < 1$. It is therefore impossible to satisfy both inequalities in (A4.5). So f must be irreducible.

It is now clear that the polynomial $x^5 + 7x^4 - 3x^2 + x + 1$, mentioned above, is irreducible. It has integer coefficients, leading coefficient 1 and non-zero constant term, and we have seen that in this case (A4.2) can be satisfied with $R = 1$. The other polynomial $2x^4 + 5x^3 + x + 1$ also has integer coefficients, non-zero constant term, and (A4.2) can be satisfied with $R = 1$ (indeed with a slightly smaller R), but its leading coefficient is 2. So we cannot conclude that it is irreducible, and actually $2x^4 + 5x^3 + x + 1 = (2x + 1)(x^3 + 2x^2 - x + 1)$.

We can strengthen the test to include some polynomials with larger leading coefficients if, at the same time, we insist on (A4.2) being satisfied with a value of R significantly less than 1. Consider a polynomial f that has integer coefficients, non-zero constant term and non-zero leading coefficient a_n, and suppose that we can satisfy (A4.2) with $R = \frac{1}{|a_n|}$. Then, if $\alpha_1, \alpha_2, \ldots, \alpha_n$ are all the roots of f, we can write

$$f(x) = a_n(x - \alpha_1) \ldots (x - \alpha_n) \tag{A4.6}$$

and if f were composite we would have

$$(b_r x^r + \ldots + b_0)(c_s x^s + \ldots + c_0) = a_n(x - \alpha_1)\ldots(x - \alpha_n) \quad \text{(A4.7)}$$

where $r \geqslant 1$, $s \geqslant 1$, $r + s = n$ and $b_r c_s = a_n$. As before, we could relabel the roots so that $b_r x^r + \ldots + b_0 = b_r(x - \alpha_1)\ldots(x - \alpha_r)$ and $c_s x^s + \ldots + c_0 = c_s(x - \alpha_{r+1})\ldots(x - \alpha_n)$, and then equating constant terms would give

$$|\alpha_1 \ldots \alpha_r| = \frac{|b_0|}{|b_r|} \geqslant \frac{1}{|b_r|} \geqslant \frac{1}{|a_n|} \quad \text{and}$$

$$|\alpha_{r+1} \ldots \alpha_n| = \frac{|c_0|}{|c_s|} \geqslant \frac{1}{|c_s|} \geqslant \frac{1}{|a_n|}. \quad \text{(A4.8)}$$

In this case, Rouché's theorem says that only one of the roots has modulus greater than or equal to $\frac{1}{|a_n|}$. So in particular either each modulus of $\alpha_1, \ldots, \alpha_r$ is less than $\frac{1}{|a_n|}$ or each modulus of $\alpha_{r+1}, \ldots, \alpha_n$ is less than $\frac{1}{|a_n|}$. Thus we cannot have both $|\alpha_1 \ldots \alpha_r| \geqslant \frac{1}{|a_n|}$ and $|\alpha_{r+1} \ldots \alpha_n| \geqslant \frac{1}{|a_n|}$, as (A4.8) would require for some $r \geqslant 1$ if f were composite. It follows that f must be irreducible.

The strengthened form of the test shows for instance that $2x^5 - 27x^4 + x^2 + 1$ is irreducible. It has integer coefficients with both constant term and leading coefficient being non-zero, and when $|x| = \frac{1}{2}$ we have

$$\left|2x^5 + x^2 + 1\right| \leqslant \left|2x^5\right| + \left|x^2\right| + 1 = \frac{21}{16} < \left|-27x^4\right| = \frac{27}{16}.$$

Similarly $3x^4 + 70x^3 - x + 2$ is irreducible because, for $|x| = \frac{1}{3}$,

$$\left|3x^4 - x + 2\right| \leqslant \left|3x^4\right| + |x| + 2 = \frac{64}{27} < \left|70x^3\right| = \frac{70}{27}.$$

〉 Appendix 5

〉 Dirichlet's theorem and Pell's equation

In this Appendix we shall study the equation $a^2 - Db^2 = 1$ and show that when D is a positive integer, other than a square, it can be satisfied by infinitely many distinct pairs of integers a, b. First we need a theorem due to G L Dirichlet in 1842 about the approximation of real numbers by rationals. Suppose that θ is a real number and $Q > 1$ is an integer. Then Dirichlet's result is that there are always integers p, q, with $0 < q < Q$, which satisfy

$$|q\theta - p| \leqslant \frac{1}{Q}. \tag{A5.1}$$

For example it is easy to see that this is true when $Q = 2$. We pick $q = 1$ (the only positive value less than 2) and then to satisfy (A5.1) we would want to choose an integer p with $|\theta - p| \leqslant \frac{1}{2}$. We can do that because the number θ is between two consecutive integers and one of them would be a suitable choice of p. In other words, whatever the value of θ, there will be some integer whose distance from θ is at most $\frac{1}{2}$. The general theorem says further that there is a multiple $q'\theta$ of θ which is at most $\frac{1}{10}$ away from an integer, with a suitable multiplier q' less than 10; there is a multiple $q''\theta$ which is at most $\frac{1}{100}$ away from an integer, with q'' less than 100; and similarly there will be multiples of θ as close to integers as we like.

We begin by looking at the multiples $1\theta = \theta, 2\theta, \ldots, (Q-1)\theta$, and for each multiplier i with $1 \leqslant i \leqslant Q - 1$ we let p_i be the largest integer not exceeding the multiple $i\theta$, so that $0 \leqslant i\theta - p_i < 1$. Now consider the $Q + 1$ numbers

$$0, \theta - p_1, 2\theta - p_2, \ldots, (Q-1)\theta - p_{Q-1} \text{ and } 1. \tag{A5.2}$$

Notice that numbers 0 and 1 here can also be written in the form $i\theta - p'$ by taking $i = 0$ and $p' = 0$ or -1. All the numbers lie in the interval $[0,1]$ and we divide that interval into the following Q subintervals:

$$[0, \frac{1}{Q}), [\frac{1}{Q}, \frac{2}{Q}), \dots, [\frac{Q-1}{Q}, 1]. \tag{A5.3}$$

The $Q+1$ numbers in (A5.2) must all be distributed among the Q intervals in (A5.3). So, by Dirichlet's frequently used 'box principle', at least one of these subintervals must contain two or more of the numbers (because there are not enough subintervals for each to contain just one number). If say $i\theta - p'$ and $j\theta - p''$ are two numbers in the same subinterval in (A5.3), then

$$\left|(i\theta - p') - (j\theta - p'')\right| \leqslant \frac{1}{Q} \tag{A5.4}$$

as every subinterval is of length $\frac{1}{Q}$. Also i and j are both between 0 and $Q - 1$ inclusive, and they must be distinct because no subinterval contains both 0 and 1 (the only two numbers in the list (A5.2) with the same multiplier). We may suppose the names are such that $i > j$, so that $Q > i > j \geqslant 0$. We then put $q = i - j$ and $p = p' - p''$ and have

$$|q\theta - p| \leqslant \frac{1}{Q} \quad \text{and} \quad 0 < q < Q \tag{A5.5}$$

as required.

For example if $\theta = \frac{374}{613}$ and $Q = 10$ then $q = 5$ and $p = 3$ give a q less than 10 with $|q\theta - p| \leqslant \frac{1}{10}$; and a choice of $q < 100$ making $q(\frac{374}{613})$ within $\frac{1}{100}$ of an integer is $q = 59$, since $\left|59(\frac{374}{613}) - 36\right| = 0.0032\ldots < \frac{1}{100}$. Similarly if $\theta = \sqrt{3}$ we can have $\left|q\sqrt{3} - p\right| \leqslant \frac{1}{10}$ with $q = 4$, $p = 7$; and $\left|56\sqrt{3} - 97\right| = 0.005\ldots$ so that $q = 56$ and $p = 97$ satisfy (A5.5) when $\theta = \sqrt{3}$ and $Q = 100$. Note that if p and q satisfy (A5.5) they must also satisfy

$$\left|\theta - \frac{p}{q}\right| \leqslant \frac{1}{qQ} < \frac{1}{q^2} \tag{A5.6}$$

since $\frac{1}{qQ} < \frac{1}{q^2}$ follows from $q < Q$. So the fraction $\frac{p}{q}$ will be a very close approximation to θ. In the above examples the fractions $\frac{3}{5}$ and $\frac{36}{59}$

are very good approximations to $\frac{374}{613}$ and the fractions $\frac{7}{4}$ and $\frac{97}{56}$ are very good approximations to $\sqrt{3}$. (For instance from (A5.6) we can see that the difference between $\sqrt{3}$ and $\frac{97}{56}$ is less than $1/56^2 = 0.0003\ldots$.)

There is a significant difference here between examples with rational values of θ and those with irrational values of θ. For $\theta = \frac{374}{613}$ we have seen how (A5.5) can be satisfied with $Q = 10$ or $Q = 100$. When $Q = 1000$ the only value of $q < 1000$ that makes $q(\frac{374}{613})$ differ from an integer by at most $\frac{1}{1000}$ is $q = 613$; and then $613(\frac{374}{613})$ is actually equal to the integer 374, so that $|613(\frac{374}{613}) - 374| = 0$. There is no point in looking for further values of q since we certainly can not make $|q(\frac{374}{613}) - p|$ smaller than zero, although multiples of $q = 613$ will match this difference (such as $q = 1226$, $p = 748$). This is quite typical of any fraction. After a certain point the values of p and q that satisfy (A5.5) will just be multiples of the fraction's numerator and denominator. It is not difficult to see why this is so. Suppose that the fraction in question is $\theta = \frac{u}{v}$ and that for a certain Q we have found $q < Q$, and an associated p, which satisfy (A5.5) and also make $|q(\frac{u}{v}) - p|$ non-zero. Then $0 < |q(\frac{u}{v}) - p| \leqslant \frac{1}{Q}$ and, as in (A5.6), we will have

$$0 < \left| \frac{u}{v} - \frac{p}{q} \right| \leqslant \frac{1}{qQ}. \tag{A5.7}$$

Now $\left| \frac{u}{v} - \frac{p}{q} \right| = \left| \frac{uq-vp}{vq} \right| = \frac{|uq-vp|}{vq}$ is a fraction with denominator vq, and it is non-zero, so the smallest it can be is $\frac{1}{vq}$. Therefore (A5.7) implies $\frac{1}{vq} \leqslant \frac{1}{qQ}$ and that in turn means $Q \leqslant v$. In other words, when $\theta = \frac{u}{v}$ and $Q > v$, the inequality (A5.5) can only be satisfied with $|q(\frac{u}{v}) - p| = 0$, which is the same as $\frac{u}{v} = \frac{p}{q}$.

The situation is quite different when θ is irrational since such a θ cannot be equal to any $\frac{p}{q}$, which is to say that $|q\theta - p|$ can never be zero. So no matter how large we choose Q (and how small $\frac{1}{Q}$) Dirichlet's theorem assures us that there will be values of p and q satisfying

$$0 < |q\theta - p| \leqslant \frac{1}{Q} \quad \text{and} \quad 0 < q < Q. \tag{A5.8}$$

When $\theta = \sqrt{3}$ and $Q = 10, 100$ we have already seen that (A5.8) can be satisfied by choosing the pair (p, q) to be $(7, 4), (97, 56)$ respectively;

with $\frac{7}{4}$, $\frac{97}{56}$ as corresponding fractions that are close to $\sqrt{3}$. Both $\left|4\sqrt{3} - 7\right|$ and $\left|56\sqrt{3} - 97\right|$ are greater than $\frac{1}{1000}$, so neither of those pairs can satisfy (A5.8) with $Q = 1000$. But if $Q = 1000$ we can take $p = 1351$, $q = 780$, giving a new fraction $\frac{1351}{780}$ that is very close to $\sqrt{3}$. Then to establish the existence of yet another good approximation to $\sqrt{3}$ we take a further value of Q large enough so that $\frac{1}{Q}$ is less than the non-zero numbers $\left|4\sqrt{3} - 7\right|$, $\left|56\sqrt{3} - 97\right|$ and $\left|780\sqrt{3} - 1351\right|$. It happens that $Q = 10\,000$ will do, and then (A5.8) can be satisfied by choosing $p = 5042$, $q = 2911$. This process can be continued for as long as we like, giving an unending sequence of different pairs (p, q) that satisfy (A5.8) for larger and larger values of Q and a corresponding sequence of fractions $\frac{p}{q}$ that each satisfy (A5.6). Our example has been with $\theta = \sqrt{3}$, but the method is exactly the same for any irrational θ. That is to say, suppose that we have constructed a sequence of different pairs (p_1, q_1), (p_2, q_2), ..., (p_n, q_n) that each satisfy (A5.8) for some values of Q, say $Q_1, Q_2, \ldots, Q_n$. Then we choose a new limit Q_{n+1} so large that $1/Q_{n+1}$ is less than each of $|q_1\theta - p_1|$, $|q_2\theta - p_2|$, ..., $|q_n\theta - p_n|$ and apply Dirichlet's result again. That will give a pair of integers p_{n+1}, q_{n+1} that satisfy $q_{n+1} < Q_{n+1}$ and $|q_{n+1}\theta - p_{n+1}| \leqslant \frac{1}{Q_{n+1}}$; and the fraction $\frac{p_{n+1}}{q_{n+1}}$ cannot be equal to any of the previous fractions since, for each $i = 1, \ldots, n$

$$|q_{n+1}\theta - p_{n+1}| \leqslant \frac{1}{Q_{n+1}} < |q_i\theta - p_i|.$$

This process of finding new pairs (p_i, q_i) can be continued indefinitely, and in every case we have

$$0 < |q_i\theta - p_i| \leqslant \frac{1}{Q_i} \quad \text{and} \quad 0 < q_i < Q_i. \tag{A5.9}$$

As in (A5.6), each of these pairs (p_i, q_i) satisfies $\left|\theta - \frac{p_i}{q_i}\right| \leqslant \frac{1}{q_i Q_i} < \frac{1}{q_i^2}$. So for any irrational number θ there must be infinitely many fractions $\frac{p}{q}$ satisfying

$$\left|\theta - \frac{p}{q}\right| < \frac{1}{q^2}. \tag{A5.10}$$

Further details about fractions $\frac{p}{q}$ satisfying (A5.10) and methods of constructing them can be found in Jackson(1987).

We are now ready to investigate the equation

$$a^2 - Db^2 = 1 \qquad \text{(A5.11)}$$

where D is a rational integer and we want to see how many integers a, b can be solutions. Of course if D is negative or zero, the left-hand side of the equation is the sum of two non-negative integers and the only integer solutions are $a = \pm 1$, $b = 0$ and, if $D = -1$, $a = 0$, $b = \pm 1$. If D is positive and a square the left side is the difference of two squares, and the smallest such a difference can be is when the squares are consecutive, which again leads to $a = \pm 1$, $b = 0$ as the only possibilities. So the only difficult cases are when D is positive and not a square. It was Fermat who in 1657 stated that there are infinitely many pairs of integers a, b that satisfy (A5.11) in these cases, but the equation is commonly known as Pell's equation due to a later misunderstanding by Euler. The proof we shall give is due to R Dedekind and G Dirichlet in the middle of the 19th century.

Our starting point is the fact that when D is positive and not a square then $\sqrt{D}$ is irrational. One way to see that is to note that D can be expressed as $k^2 d$ where $d > 1$ is square-free (has no square divisors bigger than 1). Then $\sqrt{D} = k\sqrt{d}$ so that $\sqrt{D}$ is irrational if and only if $\sqrt{d}$ is. But d must have at least one prime divisor and, using it, Eisenstein's criterion implies that the polynomial $x^2 - d$ is irreducible in $Q[x]$, which is the same as saying that $\sqrt{d}$ is irrational. Another interesting way of proving that $\sqrt{D}$ is irrational when D is not a square is given in Estermann(1975). So the result in (A5.10) implies that there will be infinitely many distinct pairs of integers x, y, with $y > 0$, such that

$$\left| \sqrt{D} - \frac{x}{y} \right| < \frac{1}{y^2}.$$

Each of these pairs then satisfies

$$\left| x - y\sqrt{D} \right| < \frac{1}{y}$$

$$\text{and} \quad \left| x + y\sqrt{D} \right| \leqslant \left| (x - y\sqrt{D}) \right| + \left| 2y\sqrt{D} \right| < 2y\sqrt{D} + \frac{1}{y}$$

whence

$$\left| x^2 - Dy^2 \right| = \left| x - y\sqrt{D} \right| * \left| x + y\sqrt{D} \right| < 2\sqrt{D} + \frac{1}{y^2} \leqslant 2\sqrt{D} + 1.$$

$$\text{(A5.12)}$$

None of these values of $x^2 - Dy^2$ can be zero (since $\sqrt{D}$ is irrational); so there are infinitely many pairs (x, y) associated with the finitely many non-zero integers lying between $-2\sqrt{D} - 1$ and $2\sqrt{D} + 1$. It follows that there must be some non-zero integer m in this range that is the value of infinitely many of the expressions $x^2 - Dy^2$. So there is an equation

$$x^2 - Dy^2 = m \qquad (A5.13)$$

with $m \neq 0$, that has infinitely many different integer solution pairs (x, y); and, by changing signs if necessary, we can still have infinitely many solutions with each of x, y positive. Now there are only finitely many possibilities for the residues of x, y modulo $|m|$, so as before there must be infinitely many different pairs that have the same residues. In particular there will be distinct pairs of positive integers (x_1, y_1) and (x_2, y_2) satisfying

$$x_1^2 - Dy_1^2 = m, \quad x_2^2 - Dy_2^2 = m, \quad x_1 \equiv x_2 (\text{mod } |m|), \quad y_1 \equiv y_2 \ (\text{mod } |m|).$$
$$(A5.14)$$

Note that $(x_1, y_1) \neq (x_2, y_2)$ means that either or both of $x_1 \neq x_2$, $y_1 \neq y_2$ hold. So we cannot have $y_1 = y_2$, for then (A5.14) would imply $x_1 = x_2$ also. This in turn means that

$$\frac{x_2}{x_1} \neq \frac{y_2}{y_1} \quad \text{and so} \quad x_2 y_1 - x_1 y_2 \neq 0; \qquad (A5.15)$$

for otherwise

$$m = x_2^2 - Dy_2^2 = \left(\frac{y_2}{y_1}\right)^2 (x_1^2 - Dy_1^2) = m\left(\frac{y_2}{y_1}\right)^2$$

contradicting $y_1 \neq y_2$. However the congruences in (A5.14) certainly imply

$$x_2 y_1 - x_1 y_2 \equiv 0 \ (\text{mod } |m|) \qquad (A5.16)$$

and

$$x_2 x_1 - Dy_2 y_1 \equiv x_1^2 - Dy_1^2 \equiv 0 \ (\text{mod } |m|). \qquad (A5.17)$$

Therefore if we define t, u by

$$t = |x_2 x_1 - Dy_2 y_1|/|m| \quad u = |x_2 y_1 - x_1 y_2|/|m| \qquad (A5.18)$$

we see that t and u are integers with $u > 0$. It turns out that $t^2 - Du^2 = 1$ and, to derive this, we note that

$$(x_1 + y_1\sqrt{D}) * (x_2 - y_2\sqrt{D}) = (x_2x_1 - Dy_2y_1) + \sqrt{D}(x_2y_1 - x_1y_2)$$

and

$$(x_1 - y_1\sqrt{D}) * (x_2 + y_2\sqrt{D}) = (x_2x_1 - Dy_2y_1) - \sqrt{D}(x_2y_1 - x_1y_2).$$

Multiplying these two identities together gives

$$(x_1^2 - Dy_1^2)(x_2^2 - Dy_2^2) = (x_2x_1 - Dy_2y_1)^2 - D(x_2y_1 - x_1y_2)^2$$

or

$$m^2 = m^2t^2 - Dm^2u^2$$

which is

$$t^2 - Du^2 = 1. \tag{A5.19}$$

This may seem like a lot of needless work in order to get just one solution $a = t$, $b = u$ of (A5.11); especially when we already know a solution: $a = 1, b = 0$. But the point is that we now know a solution with $b \neq 0$, and from this it is not hard to make up infinitely many solutions. First of all note that any solution of $a^2 - Db^2 = 1$ with $b \neq 0$ must also have $a \neq 0$ and, by changing signs if necessary, we may assume that a and b are both positive. Since we now know that there are some integer solutions like this, there will be one, say $a = t_1 > 0$, $b = u_1 > 0$, with the smallest possible $b > 0$ out of all the integer solutions. The solution (t_1, u_1) is called the *fundamental solution* of (A5.11). For each positive integer n we can construct another solution (t_n, u_n) in terms of t_1, u_1 by defining

$$t_n + u_n\sqrt{D} = (t_1 + u_1\sqrt{D})^n \quad \text{and} \quad t_n - u_n\sqrt{D} = (t_1 - u_1\sqrt{D})^n. \tag{A5.20}$$

Then $a = t_n$, $b = u_n$ is a solution of $a^2 - Db^2 = 1$ because

$$t_n^2 - Du_n^2 = (t_1 + u_1\sqrt{D})^n(t_1 - u_1\sqrt{D})^n = (t_1^2 - Du_1^2)^n = 1. \tag{A5.21}$$

It is also easy to see that (t_n, u_n) is an integer solution. For example, since

$$t_{n+1} + u_{n+1}\sqrt{D} = (t_n + u_n\sqrt{D})(t_1 + u_1\sqrt{D})$$

$$\text{and} \quad t_{n+1} - u_{n+1}\sqrt{D} = (t_n - u_n\sqrt{D})(t_1 - u_1\sqrt{D})$$

we have $t_{n+1} = t_n t_1 + D u_n u_1$ and $u_{n+1} = t_n u_1 + u_n t_1$ so that inductively if t_1, u_1, t_n, u_n are all integers then so are t_{n+1}, u_{n+1}. Further, from (A5.20) the numbers $t_n + u_n\sqrt{D}$ are increasing, so the pairs (t_n, u_n) are all different.

We have now done what we set out to do; namely show that when D is positive and not a square there are infinitely many integer pairs a, b that satisfy the equation $a^2 - Db^2 = 1$. We did this by using Dirichlet's general approximation result to show that there is at least one solution with b positive. Then from the solution $a = t_1 > 0$, $b = u_1 > 0$, with the smallest positive b, (A5.20) gave infinitely many solutions. We shall complete the theory by showing that (A5.20) gives all the positive solutions (so that every solution is of the form $a = \pm t_n$, $b = \pm u_n$ for some n). So suppose that $a = t$, $b = u$ is a solution with t and u both positive integers. Then $u \geqslant u_1$ and, from $t^2 - Du^2 = t_1^2 - Du_1^2$, we will also have $t \geqslant t_1$ whence $t + u\sqrt{D} \geqslant t_1 + u_1\sqrt{D}$. So there will be a positive integer n such that

$$t_n + u_n\sqrt{D} = (t_1 + u_1\sqrt{D})^n \leqslant t + u\sqrt{D} < (t_1 + u_1\sqrt{D})^{n+1}. \quad \text{(A5.22)}$$

Now define T and U by

$$T + U\sqrt{D} = (t + u\sqrt{D})(t_n - u_n\sqrt{D})$$

$$\text{(A5.23)}$$

$$\text{and} \quad T - U\sqrt{D} = (t - u\sqrt{D})(t_n + u_n\sqrt{D}).$$

This makes $T = tt_n - Duu_n$ and $U = ut_n - tu_n$ both integers and

$$T^2 - DU^2 = (t^2 - Du^2)(t_n^2 - Du_n^2) = 1. \quad \text{(A5.24)}$$

Also $(t_n + u_n\sqrt{D})(t_n - u_n\sqrt{D}) = 1$ means that $t_n - u_n\sqrt{D}$ is the reciprocal of $t_n + u_n\sqrt{D}$ so, from (A5.23), $T + U\sqrt{D}$ is therefore $(t + u\sqrt{D})/(t_n + u_n\sqrt{D})$. Hence, from (A5.22),

$$1 \leqslant T + U\sqrt{D} < t_1 + u_1\sqrt{D}. \quad \text{(A5.25)}$$

As before, $T - U\sqrt{D}$ is $(T + U\sqrt{D})^{-1}$, so that $0 < T - U\sqrt{D} \leqslant 1$ and therefore

$$T = \frac{1}{2}[(T + U\sqrt{D}) + (T - U\sqrt{D})] > 0$$

$$\text{and} \quad U = \frac{1}{2\sqrt{D}}[(T + U\sqrt{D}) - (T - U\sqrt{D})] \geqslant 0.$$

If now U were non-zero, then T and U would both be positive and necessarily $U \geqslant u_1$ since u_1 is the smallest positive value of b in $a^2 - Db^2 = 1$. From $T^2 - DU^2 = t_1^2 - Du_1^2$ we would also have to have $T \geqslant t_1$; but then (A5.25) would not be satisfied. So U has to be zero and (A5.24) implies that T is 1. Thus, from (A5.23),

$$t + u\sqrt{D} = (t_n - u_n\sqrt{D})^{-1} = t_n + u_n\sqrt{D}$$

$$\text{and} \quad t - u\sqrt{D} = (t_n + u_n\sqrt{D})^{-1} = t_n - u_n\sqrt{D}$$

giving $t = t_n$, $u = u_n$ as we wanted.

For example we now know that the equation $a^2 - 12b^2 = 1$ has solutions other than $a = \pm 1$, $b = 0$. In this case the smallest solution with a and b both positive is $t_1 = 7$, $u_1 = 2$. Following (A5.20), the next positive solution is then obtained from $t_2 + u_2\sqrt{12} = (7 + 2\sqrt{12})^2$ and $t_2 - u_2\sqrt{12} = (7 - 2\sqrt{12})^2$ giving $t_2 = 97$, $u_2 = 28$; and the one after that from $t_3 + u_3\sqrt{12} = (7 + 2\sqrt{12})^3$, $t_3 - u_3\sqrt{12} = (7 - 2\sqrt{12})^3$ making $t_3 = 18817$, $u_3 = 5432$.

Note that, from (A5.20), we have

$$t_n - u_n\sqrt{D} = [1/(t_1 + u_1\sqrt{D})]^n = (t_1 + u_1\sqrt{D})^{-n}$$

$$\text{(A5.26)}$$

$$t_n + u_n\sqrt{D} = (t_1 - u_1\sqrt{D})^{-n}.$$

In other words the solution $(t_n, -u_n)$ can be obtained from (A5.20) by changing the sign of n. Thus every solution (t, u) of $a^2 - Db^2 = 1$ is determined by

$$t + u\sqrt{D} = \pm(t_1 + u_1\sqrt{D})^m \quad \text{for } m \in \mathbb{Z} \qquad \text{(A5.27)}$$

and its conjugate equation with $\sqrt{D}$ replaced by $-\sqrt{D}$.

We can also use our analysis to draw conclusions about some other similar equations. When D is positive and not a square the equation

$$a^2 - Db^2 = 4 \qquad \text{(A5.28)}$$

for instance will have infinitely many solutions. This is because if $t_n^2 - Du_n^2 = 1$ then $(2t_n)^2 - D(2u_n)^2 = 4$, and so the infinitely many solutions of $a^2 - Db^2 = 1$ provide infinitely many solutions of (A5.28). Thus from the solutions $(7, 2)$ and $(97, 28)$ of $a^2 - 12b^2 = 1$ we get the solutions $(14, 4)$ and $(194, 56)$ of $a^2 - 12b^2 = 4$. In this case there are other solutions, such as $(4, 1)$, as well as those obtained by doubling the solutions of $a^2 - 12b^2 = 1$. Defining the fundamental solution $a = t_1 > 0$, $b = u_1 > 0$ of the equation $a^2 - Db^2 = 4$, as before, to be the solution with a and b both positive and smallest possible b, then a similar analysis to that above shows that all the positive solutions t_n, u_n of (A5.28) are given by

$$\tfrac{1}{2}(t_n + u_n\sqrt{D}) = \left[\tfrac{1}{2}(t_1 + u_1\sqrt{D})\right]^n$$

(A5.29)

$$\text{and} \quad \tfrac{1}{2}(t_n - u_n\sqrt{D}) = \left[\tfrac{1}{2}(t_1 - u_1\sqrt{D})\right]^n.$$

For $D = 12$, the smallest solution is $(4, 1)$ and so the next solution of $a^2 - 12b^2 = 4$ is (t_2, u_2) where $\tfrac{1}{2}(t_2 + u_2\sqrt{12}) = \left[\tfrac{1}{2}(4 + \sqrt{12})\right]^2$ and $\tfrac{1}{2}(t_2 - u_2\sqrt{12}) = \left[\tfrac{1}{2}(4 - \sqrt{12})\right]^2$, leading to $t_2 = 14$, $u_2 = 4$ which we have already noticed. The third positive solution is given by $\tfrac{1}{2}(t_3 + u_3\sqrt{12}) = \left[\tfrac{1}{2}(4 + \sqrt{12})\right]^3$ and $\tfrac{1}{2}(t_3 - u_3\sqrt{12}) = \left[\tfrac{1}{2}(4 - \sqrt{12})\right]^3$, which makes $t_3 = 52$, $u_3 = 15$; and $(194, 56)$ turns out to be the fourth solution. As in (A5.27) all solutions of (A5.28) are given by

$$\tfrac{1}{2}(t + u\sqrt{D}) = \pm\left[\tfrac{1}{2}(t_1 + u_1\sqrt{D})\right]^m$$

(A5.30)

where m ranges over all rational integers.

The equations

$$a^2 - Db^2 = -1$$

(A5.31)

and

$$a^2 - Db^2 = -4$$

(A5.32)

also occur naturally when considering units in quadratic domains (see (4.7) and (4.8) in Chapter 4); but there are many cases in which they do not have solutions. Of course they cannot be satisfied when D is negative or zero; and when D is a square the left-hand side of each is the difference of two squares and so $a = 0$ and $Db^2 = 1$ or 4, making the only possibilities $D = 1$, $b = \pm 1$ in the first case, $D = 1$, $b = \pm 2$

and $D = 4$, $b = \pm 1$ in the second. Even when D is positive and not a square there are still values of D for which there are no solutions. For instance neither of the equations $a^2 - 12b^2 = -1$, $a^2 - 12b^2 = -4$ have integer solutions because, modulo 3, they each reduce to $a^2 \equiv -1$ (mod 3) which does not hold for any integer a. However when D is positive and not a square and the equations have any solutions they will have infinitely many.

For suppose that (A5.31) has solutions for some positive non-square D and that its fundamental solution is $(a, b) = (r_1, s_1)$. That is,

$$r_1^2 - Ds_1^2 = -1 \tag{A5.33}$$

and s_1 is the least positive value of b out of all the solutions of (A5.31). Then, as in (A5.20) and (A5.21), the sequence of pairs (r_{2k-1}, s_{2k-1}) defined for each $k \geqslant 1$ by

$$r_{2k-1} + s_{2k-1}\sqrt{D} = (r_1 + s_1\sqrt{D})^{2k-1}$$

$$r_{2k-1} - s_{2k-1}\sqrt{D} = (r_1 - s_1\sqrt{D})^{2k-1} \tag{A5.34}$$

is an infinite sequence of pairs of integers satisfying (A5.31); and the pairs (r_{2k}, s_{2k}) defined similarly by

$$r_{2k} + s_{2k}\sqrt{D} = (r_1 + s_1\sqrt{D})^{2k}$$

$$r_{2k} - s_{2k}\sqrt{D} = (r_1 - s_1\sqrt{D})^{2k} \tag{A5.35}$$

all satisfy $r_{2k}^2 - Ds_{2k}^2 = 1$. An analysis similar to the discussion beginning at (A5.22) shows that the fundamental solution (t_1, u_1) of $a^2 - Db^2 = 1$ is actually given in terms of (r_1, s_1) by

$$t_1 + u_1\sqrt{D} = (r_1 + s_1\sqrt{D})^2 \quad t_1 - u_1\sqrt{D} = (r_1 - s_1\sqrt{D})^2$$

and all the positive solutions of (A5.11) are given by the pairs (r_{2k}, s_{2k}) in (A5.35). Also all the positive solutions of (A5.31) are given by the pairs (r_{2k-1}, s_{2k-1}) in (A5.34). Hence every pair (r, s) that satisfies either (A5.11) or (A5.31) is given by

$$r + s\sqrt{D} = \pm(r_1 + s_1\sqrt{D})^m \quad \text{for } m \in \mathbb{Z}. \tag{A5.36}$$

Similarly if (A5.32) has solutions for some positive D that is not a square and if $(a, b) = (r_1, s_1)$ is its least positive solution, then the pairs (r_n, s_n) defined by

$$\tfrac{1}{2}(r_n + s_n\sqrt{D}) = \left[\tfrac{1}{2}(r_1 + s_1\sqrt{D})\right]^n$$

and $\quad \tfrac{1}{2}(r_n - s_n\sqrt{D}) = \left[\tfrac{1}{2}(r_1 - s_1\sqrt{D})\right]^n \qquad$ (A5.37)

give all the positive solutions of (A5.32) if n is odd, and all positive solutions of (A5.28) if n is even. So, as in (A5.36), the pairs determined by

$$\tfrac{1}{2}(r + s\sqrt{D}) = \pm\left[\tfrac{1}{2}(r_1 + s_1\sqrt{D})\right]^m \qquad \text{for } m \in \mathbb{Z}. \qquad \text{(A5.38)}$$

give all solutions of both (A5.28) and (A5.32).

⟩ Appendix 6

⟩ Quadratic reciprocity

Legendre's symbol has a very powerful property which, as we shall see, enables the value of $\left(\frac{a}{p}\right)$ to be calculated with hardly any trouble, at least for modest values of the arguments a and p. When p and q are different odd primes the property relates $\left(\frac{p}{q}\right)$ and $\left(\frac{q}{p}\right)$ and enables the value of either one to be written down once the other is known. This 'reciprocity' property is that

$$\left(\frac{p}{q}\right)\left(\frac{q}{p}\right) = (-1)^{\left(\frac{p-1}{2}\right)\left(\frac{q-1}{2}\right)} \tag{A6.1}$$

or, multiplying both sides by $\left(\frac{q}{p}\right)$,

$$\left(\frac{p}{q}\right) = (-1)^{\left(\frac{p-1}{2}\right)\left(\frac{q-1}{2}\right)}\left(\frac{q}{p}\right) \tag{A6.2}$$

since $\left(\frac{q}{p}\right)^2$ is certainly 1. In (A6.2) the exponent $\left(\frac{p-1}{2}\right)\left(\frac{q-1}{2}\right)$ is even if either p or q is congruent to 1 modulo 4, and is odd if both p and q are congruent to -1 modulo 4. So (A6.2) is equivalent to the statement that

$$\left(\frac{p}{q}\right) = \begin{cases} \left(\frac{q}{p}\right) & \text{if } p \equiv 1 \pmod{4} \text{ or } q \equiv 1 \pmod{4} \\ -\left(\frac{q}{p}\right) & \text{if } p \equiv q \equiv -1 \pmod{4}. \end{cases} \tag{A6.3}$$

This remarkable property was first noticed by Euler about 1745 and Legendre in 1785 and was first proved by Gauss in 1796. The proof that we shall give is due to J. S. Frame in 1978. It will be convenient to denote $\left(\frac{p-1}{2}\right)$ by p' and $\left(\frac{q-1}{2}\right)$ by q' since we frequently refer to these quantities. In the proof we shall make use of Euler's criterion (see (5.20)), which in this context says that $\left(\frac{p}{q}\right) \equiv p^{q'}$ (mod q). The right-hand side of this last congruence is a product (of q' terms each equal to p) and we shall begin by examining the seemingly unrelated product f_{pq} where

$$f_{pq} = \prod_{k=1}^{q'} \prod_{h=1}^{p'} \frac{hq - kp}{|hq - kp|}. \tag{A6.4}$$

Now hq is not equal to kp, since q cannot divide kp as p and q are different primes and $1 \leqslant k \leqslant q'$. So $hq - kp$ is not zero and therefore $\frac{hq-kp}{|hq-kp|}$ must be either $+1$ or -1. Thus the whole product f_{pq} must be either $+1$ or -1. It will turn out that f_{pq} is congruent, modulo q, to $p^{q'}$ and so to $\left(\frac{p}{q}\right)$. This will further imply that $f_{pq} = \left(\frac{p}{q}\right)$ since they are each ± 1.

First of all the expression $\frac{hq-kp}{|hq-kp|}$ only has the value -1 when $hq - kp$ is negative, which happens when $hq < kp$ or $h < \frac{kp}{q}$. This fraction $\frac{kp}{q}$ is not an integer since, as we have seen, q does not divide kp. So there are $\left[\frac{kp}{q}\right]$ positive integers less than $\frac{kp}{q}$ (namely $1, \ldots, \left[\frac{kp}{q}\right]$) and they all lie in the range $1, \ldots, p'$ because $\frac{kp}{q} \leqslant \frac{q'p}{q} < \frac{1}{2}p$. This means that for each value of k there are $\left[\frac{kp}{q}\right]$ corresponding values of h for which $\frac{hq-kp}{|hq-kp|}$ is -1. The product of these $\left[\frac{kp}{q}\right]$ expressions that are each -1 is of course $(-1)^{\left[\frac{kp}{q}\right]}$; and that is the product of all the expressions for a fixed k (since those that are $+1$ do not contribute anything new). The result is that

$$f_{pq} = \prod_{k=1}^{q'} (-1)^{\left[\frac{kp}{q}\right]}. \tag{A6.5}$$

Since f_{pq} is ± 1 and q is odd, we have further that

$$f_{pq} = f_{pq}^q = \left[\prod_{k=1}^{q'} (-1)^{\left[\frac{kp}{q}\right]}\right]^q = \prod_{k=1}^{q'} (-1)^{q\left[\frac{kp}{q}\right]}. \tag{A6.6}$$

Now $\left[\frac{kp}{q}\right]$ is the largest integer less than $\frac{kp}{q}$, so in (A6.6) the exponent $q\left[\frac{kp}{q}\right]$ is the largest multiple of q below $q\left(\frac{kp}{q}\right) = kp$. Thus kp can be written as $q\left[\frac{kp}{q}\right]$ plus a remainder strictly between 0 and q. If the remainder is between $\frac{q+1}{2}$ and $q-1$ inclusive then kp is actually nearer to the next multiple of q, namely $q\left[\frac{kp}{q}\right]+q$, and we can write

$$kp = q(\left[\frac{kp}{q}\right]+1)-r_k \quad \text{for} \quad 0 < r_k \leqslant \frac{q-1}{2}. \qquad (A6.7)$$

Otherwise we simply have

$$kp = q\left[\frac{kp}{q}\right]+r_k \quad \text{where} \quad 0 < r_k \leqslant \frac{q-1}{2}. \qquad (A6.8)$$

Only one of these possibilities can hold for a particular k. We can however work with the equations simultaneously by combining them in the form

$$kp = q(\left[\frac{kp}{q}\right]+e_k)+(-1)^{e_k}r_k \quad \text{where } e_k = 0 \text{ } or \text{ } 1,$$

$$\text{and } 0 < r_k \leqslant \frac{q-1}{2}. \qquad (A6.9)$$

Note that each k uniquely determines the corresponding values of e_k and r_k. For instance e_k is 0 when kp is nearer to $q\left[\frac{kp}{q}\right]$ than to $q\left[\frac{kp}{q}\right]+q$, and r_k is then the remainder in (A6.8); otherwise e_k is 1 and r_k is determined by (A6.7).

The equation (A6.9) implies

$$kp \equiv (-1)^{e_k}r_k \pmod{q} \qquad (A6.10)$$

and then, for k, l both in the range $1 \ldots q'$, we have

$$r_k - r_l \equiv p[(-1)^{e_k}k - (-1)^{e_l}l]] \equiv (-1)^{e_k}p(k \pm l) \pmod{q}.$$

When $k \neq l$ we have $0 < |k \pm l| < q$ so that $r_k - r_l$ cannot be zero modulo q and in particular $r_k \neq r_l$. Thus in (A6.9) the $\frac{q-1}{2}$ different values of k give $\frac{q-1}{2}$ different values of r_k. There only are $\frac{q-1}{2}$ integers in the allowable range $1, \ldots, \frac{q-1}{2}$ for r_k; so as k goes from 1 to $\frac{q-1}{2}$ the corresponding numbers r_k fill up the same interval. Consequently,

$$\sum_{k=1}^{q'} k = \sum_{k=1}^{q'} r_k = \frac{q'(q'+1)}{2 \cdot} \qquad (A6.11)$$

and

$$\prod_{k=1}^{q'} k = \prod_{k=1}^{q'} r_k = q'!$$ (A6.12)

It also follows from (A6.10) that

$$\prod_{k=1}^{q'}(kp) \equiv \prod_{k=1}^{q'}(-1)^{e_k} r_k \pmod{q}.$$

This, with (A6.12), is

$$p^{q'}(q'!) \equiv (q'!) \prod_{k=1}^{q'}(-1)^{e_k} \pmod{q}$$

which gives

$$\left(\frac{p}{q}\right) \equiv p^{q'} \equiv \prod_{k=1}^{q'}(-1)^{e_k} \pmod{q}.$$ (A6.13)

If p and q are both odd they are each congruent to 1 modulo 2 so that (A6.9) implies

$$k \equiv \left[\frac{kp}{q}\right] + e_k + (-1)^{e_k} r_k \equiv \left[\frac{kp}{q}\right] - e_k + r_k \pmod{2}.$$ (A6.14)

So, in (A6.5),

$$\begin{aligned}
f_{pq} &= \prod_{k=1}^{q'}(-1)^{\left[\frac{kp}{q}\right]} = \prod_{k=1}^{q'}(-1)^{k-r_k+e_k} \\
&= (-1)^{\sum k - \sum r_k + \sum e_k} \\
&= (-1)^{\sum e_k} \quad \text{from (A6.11)} \\
&\equiv \left(\frac{p}{q}\right) \pmod{q} \quad \text{from (A6.13).}
\end{aligned}$$

As remarked earlier this means that $f_{pq} = \left(\frac{p}{q}\right)$. Hence

$$\left(\frac{p}{q}\right)\left(\frac{q}{p}\right) = f_{pq} f_{qp} = \prod_{k=1}^{q'} \prod_{h=1}^{p'} \frac{hq - kp}{|hq - kp|} * \frac{kp - hq}{|kp - hq|}$$

$$= \prod_{k=1}^{q'} \prod_{h=1}^{p'} (-1) = (-1)^{p'q'}$$

as claimed in (A6.1).

To decide for instance whether 7 is a square modulo 997 we can now reason that

$$\left(\frac{7}{997}\right) = \left(\frac{997}{7}\right) \quad \text{since } 997 \equiv 1 \pmod 4$$

$$= \left(\frac{3}{7}\right) \quad \text{since } 997 \equiv 3 \pmod 7$$

$$= -\left(\frac{7}{3}\right) \quad \text{since } 3 \equiv 7 \equiv -1 \pmod 4$$

$$= -\left(\frac{1}{3}\right)$$

$$= -1.$$

So there is no square congruent to 7 modulo 997. Similarly

$$\left(\frac{85}{997}\right) = \left(\frac{5}{997}\right)\left(\frac{17}{997}\right)$$

$$= \left(\frac{997}{5}\right)\left(\frac{997}{17}\right)$$

$$= \left(\frac{2}{5}\right)\left(\frac{11}{17}\right)$$

$$= -\left(\frac{17}{11}\right) \quad \text{as 2 is not a square modulo 5 and } \left(\frac{11}{17}\right) = \left(\frac{17}{11}\right)$$

$$= -\left(\frac{6}{11}\right)$$

$$= +1 \quad \text{as 6 is not a square modulo 11.}$$

So 85 is a square modulo 997 (indeed $85 \equiv 303^2 \pmod{997}$).

However we can not as yet evaluate a Legendre symbol such as $\left(\frac{46}{997}\right) = \left(\frac{2}{997}\right)\left(\frac{23}{997}\right)$ because we do not know how to deal with symbols of the form $\left(\frac{2}{q}\right)$ where q is an odd prime. [When q is small we can of course find the value of $\left(\frac{2}{q}\right)$ by testing whether any of the squares

$1^2, \ldots, \left(\frac{q-1}{2}\right)^2$ are congruent to 2 modulo q, but this is impracticable for $q = 997$.] It turns out that some of the reasoning that applied to $\left(\frac{p}{q}\right)$ when p and q were both odd can also be used when $p = 2$ and q is an odd prime, but in this case we do not need any expression like f_{pq}. Here, with $0 < k \leqslant q'$, we have $0 < 2k < q$ and, instead of (A6.9),

$$2k = q * e_k + (-1)^{e_k} r_k \quad \text{where } e_k = 0 \text{ or } 1,$$
$$\text{and } 0 < r_k \leqslant \tfrac{q-1}{2}. \tag{A6.15}$$

This implies that (A6.10) is still true when $p = 2$, namely

$$2k \equiv (-1)^{e_k} r_k \pmod{q};$$

and the same reasoning as that following (A6.10) shows that as k ranges over the interval $1, \ldots, \frac{q-1}{2}$ so does r_k (not necessarily in the same order). Thus (A6.11) and (A6.12) still hold and so does (A6.13) when $p = 2$. In this case (A6.15) implies

$$0 \equiv e_k + (-1)^{e_k} r_k \equiv e_k - r_k \pmod 2.$$

or, in other words,

$$e_k \equiv r_k \pmod 2. \tag{A6.16}$$

So from (A6.13)

$$\left(\frac{2}{q}\right) \equiv \prod_{k=1}^{q'} (-1)^{e_k} \equiv \prod_{k=1}^{q'} (-1)^{r_k} = (-1)^{\sum r_k} = (-1)^{\frac{q'(q'+1)}{2}} \tag{A6.17}$$

where we have also used (A6.16) and (A6.11). Thus

$$\left(\frac{2}{q}\right) = (-1)^{\frac{(q-1)(q+1)}{8}} = (-1)^{\frac{(q^2-1)}{8}} \tag{A6.18}$$

which is the same as

$$\left(\frac{2}{q}\right) = \begin{cases} +1 & \text{if } q \equiv \pm 1 \pmod 8 \\ -1 & \text{if } q \equiv \pm 3 \pmod 8. \end{cases} \tag{A6.19}$$

We can now complete the evaluation of $\left(\frac{46}{997}\right)$ since

$$\left(\frac{46}{997}\right) = \left(\frac{2}{997}\right)\left(\frac{23}{997}\right)$$

$$= (-1) * \left(\frac{997}{23}\right) \quad \text{as } 997 \equiv 5 \pmod 8$$

$$= -\left(\frac{8}{23}\right)$$

$$= -\left(\frac{2}{23}\right)^3$$

$$= -1 \text{ as } 23 \equiv -1 \pmod 8,$$

so 46 is not a square modulo 997.

⟩ References

Ahlfors L V 1953 *Complex Analysis* (New York: McGraw-Hill)
Allenby R B J T 1991 *Rings, Fields and Groups* Second edition (London: Edward Arnold)
Anderson D F and Pruis P 1991 *Proc. Am. Math. Soc.* **113** 933–7
Chapman S T 1992 *Am. Math. Mon.* **99** 943–45
Cohn H 1980 *Advanced Number Theory* (New York: Dover)
Cox D A 1989 *Primes of the form $x^2 + ny^2$* (New York: Wiley)
Dickson L E 1927 *Bull. Am. Math. Soc.* **33** 63–70
Estermann T 1975 *Math. Gazette* **59** 110
Frame J S 1978 *Am. Math. Mon.* **85** 818–9
Grosswald E 1985 *Representations of Integers as Sums of Squares* (New York: Springer)
Jackson T H 1987 *From Number Theory to Secret Codes* (Bristol: Adam Hilger)
Lidl R and Niederreiter H 1986 *Introduction to Finite Fields and their Applications* (Cambridge: Cambridge University Press)
Ramanujan S 1917 *Proc. Camb. Phil. Soc.* **19** 11–21 (reprinted 1927 *Collected Papers of Srinivasa Ramanujan* (Cambridge: Cambridge University Press)
Serret J A 1928 *Cours D'Algébre Superieure* Septième edition, Tome Second (Paris: Gauthier-Villars)
Stark H M 1979 *An Introduction to Number Theory* (Cambridge, MA: MIT Press)
Stewart I N and Tall D O 1979 *Algebraic Number Theory* (London: Chapman & Hall)
Watson G L 1960 *Integral Quadratic Forms* (Cambridge: Cambridge University Press)

〉 Index

Abel, 139
Algebraic number, 60*ff*
Algebraic integer, *see* integer
   algebraic
Associate, 72, 143
Associative, 5, 7, 37, 39, 41, 43,
   138, 141

Cancellation law, 7, 65, 142
Common divisors
   of polynomials, 13*ff*
   in $\mathbb{Z}[i]$, 71
   in $\mathbb{Q}(\sqrt{d})$, 97, 98
   in integral domains, 143, 147
Commutative, 5, 7, 37, 39, 41, 43,
   64, 139, 140, 141
Composite
   polynomials, 17
   Gaussian integers, 72, 73, 75*ff*
   quadratic integers, 72, 73, 75*ff*,
     98*ff*, 108*ff*
   elements in integral domains,
     143
   elements in Euclidean domains,
     146*ff*
Conjugate
   algebraic, 62, 84, 87, 92
   complex, 76

Dedekind, 166
Derivative, 50*ff*, 57
Dirichlet, 162*ff*
Distributive law, 8, 38, 39, 41, 43,
   53, 140
Division with remainder, 11*ff*
   among polynomials, 11, 12
   in $\mathbb{Z}[i]$, 68*ff*
   in $\mathbb{Q}(\sqrt{d})$, 93*ff*

EGA, 76
Eisenstein's criterion, 26, 27, 33,
   158
Estermann, 166
Euclid's proof, 74, 91
Euclid's algorithm
   for polynomials, 13, 14
   in $\mathbb{Z}[i]$, 71
   in $\mathbb{Q}(\sqrt{d})$, 97
Euclidean domain, 71, 93, 96 144*ff*
Euler, 116, 166, 174
Extension field, 37

Factor theorem, 18
Factorial, 75
Factorization
   in $\mathbb{Q}[x]$, 18*ff*, 27*ff*
   in $\mathbb{Z}_p[x]$, 18*ff*, 24*ff*, 30*ff*
   in $\mathbb{Z}[x]$, 22*ff*, 30*ff*